THE TUNNEL

The Story of the Channel Tunnel 1802-1994

THE TUNNEL

The Story of the Channel Tunnel 1802-1994

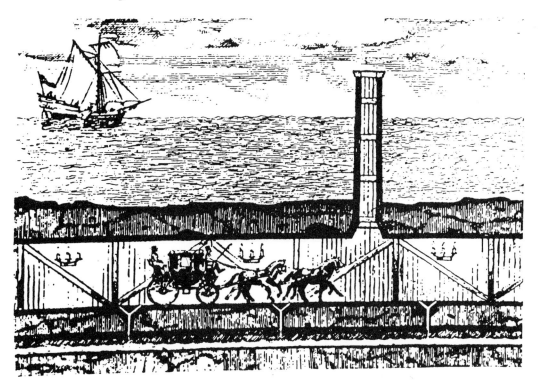

Donald Hunt

First published in Great Britain 1994 by
Images Publishing (Malvern) Ltd.,
Upton-Upon Severn,
Worcestershire.

British Library Cataloguing in Publication Data

A catalogue record for this book is available
from the British Library

ISBN 1 897817 35 7 Hardback

ISBN 1 897817 35 5 Paperback

Designed and Produced by Images Publishing (Malvern) Ltd.
Printed and Bound in Great Britain by The Bath Press, Bath, Avon.

CONTENTS

AUTHOR'S PREFACE
AND
ACKNOWLEDGEMENTS

My involvement with the Channel Tunnel began over thirty years ago, when in 1958 I became public relations consultant to the Channel Tunnel Study Group. Since then, my active association with the Tunnel has continued without interruption until the very end of 1989, and during this long period I have served a number of different masters including the British Channel Tunnel Company, British Rail, and Eurotunnel.

I swiftly became, as the French say, a *tunneliste*, and also became fascinated by the early history of the project. My research into its history, coupled with my close association with the later, day-to-day story of the events that culminated in the construction of the Tunnel, led me to write this book.

What I like to call the Anglo/French Connection is primarily about a vision, about the people who shared that vision, and about one of the greatest engineering and transport projects of the century. It is my hope that this book adequately represents these themes, and that it will be found to deal with the many branches of its subject, which include geology, engineering, politics, international relations, high finance, rail transport, and environmental and safety issues, in a way that is both balanced and accessible.

I must now make a confession. When in 1974 the second attempt to construct the Tunnel was made, a part of this book was already in draft. Shortly thereafter the British Government unilaterally abandoned it and the project became dogged by uncertainties. Yet, overall, the will to make it happen and the skills that have enabled it to happen have won through.

I have been fortunate in the help I have received during the writing of this book. First and foremost, my infinite gratitude goes to Liz, my wife, who has for twenty years shared in the drafting and taken endless pains with the many re-writes that have been necessary during the more recent history of the project. My gratitude goes in like measure to my sons Gary and Greg for their sustained support throughout. Indeed, in sharing my devotion to the Tunnel story, my family has made it all worthwhile. I am grateful to Egon Larsen for his help in verifying the early history, and to Andrew Best, for some fifteen years my friend and literary mentor, for his continuing help and guidance. I wish also to thank Tony Harold of Images Publishing who, together with his colleagues, has steered this book through to publication. They have accomplished a very tight schedule with unfailing energy and enthusiasm.

I also wish to thank those who have enabled and allowed me to reproduce certain illustrations, in particular Richard Watt for his photograph of Sir Alastair Morton and Neville Simms, The Hulton Deutsch Collection, Rex Features Ltd. and QA Photos Ltd. of Hythe, Kent, for allowing me to draw upon their archive of Tunnel-related pictures and the photograph of Eurostar used on the front cover. If inadvertently I have overlooked or been unaware of any copyright owner whose material is reproduced in these pages, I would welcome information that would enable me to set the record straight in any future printing of this book.

DONALD HUNT
London, January 1994

The book is dedicated to all those who, by their undaunted belief in the project, kept their vision of the Channel Tunnel bright during nearly two centuries of suppressed endeavour and to those, too numerous to name, who promoted and completed one of the greatest engineering achievements of the twentieth century.

PROLOGUE

What a cursed thing to live in an island! This step is more awkward than the whole journey.

Letter from Edward Gibbon to Lord Sheffield
from aboard a Channel boat, 1783.

Patriotism often fails to survive; and if any wish is felt in mid-Channel, it is that, after all, England is not an island.

Gentleman's Magazine, circa 1860.

That unspeakable horror . . . an hour and half's torture . . . two centuries of agony.

Nineteenth century descriptions of the Channel crossing.

The British have always accepted the discomforts and uncertainties of Channel crossings as the price to be paid for their splendid isolation. For a thousand years their shores have been free from foreign invasion and, as a nation of sailors and colonisers who reached out to the far corners of the earth, they were not to be put off by a trifle such as the Straits of Dover. Then, in 1909, came the day when the British press declared: 'England has ceased to be an island.' Louis Blériot had crossed the Straits of Dover in an aeroplane. Yet for nearly another hundred years, for cross-Channel travellers and traders, England was to remain very much an island.

It has not always been so. Scientists tell us that England's 'moat' is a temporary geographic feature, and that in a few thousand years, with the coming of a new Ice Age, Britain could once more be joined to the continental mainland as it has been several times in prehistory. Scientific opinion varies, but the generally accepted estimate is that England last became separated from the Continent at the end of the last Ice Age, some 9,000 years ago, when the waters flowed back gradually to complete Britain's island status.

The Channel seabed, however, is very much older. According to geologists, it took some two hundred million years to form. Primary-era rocks are its solid basis, with layers of Upper Jurassic rocks covered by vast deposits of dead marine organisms. The shells of these organisms now form several layers of calcium carbonate strata hundreds of feet thick, with up to 200 feet of rather turbulent water above.

The Channel is known to mariners as one of the world's most dangerous passages, with its tidal streams, untidy currents, and frequent thick fog. Yet the Straits of Dover, at its narrowest, is no more than 21 miles across, and for the better part of two centuries engineers and laymen have been arguing that the logical answer to the perils, inconvenience and cost of the crossing would be to travel *under* it.

The term 'tunnel' came into general usage in the early 1800s, but the subterranean, or underwater, passage has a surprisingly long history. The first structure of this kind was made under the River Euphrates in 2170 BC, at the command of a Babylonian queen to provide her with a safe passage between her palace on one side of the river and her temple on the far bank, and spanned a distance of nearly a mile. It was not a bored tunnel, but the first known example of the 'cut-and-cover' technique employed to this day. The Babylonian engineers diverted the river into a temporary bed so that an army of slaves could dig a trench 15 feet wide and 13 feet deep in the dry river bed. When completed, it was lined with sun-dried bricks, covered with earth, and the Euphrates was returned to its original course.

A number of tunnels were built by the Egyptians through solid rock, an example being a 650-foot-long, partly inclined passage through a hill at Thebes, which led to the burial chamber of the Pharaoh Mineptah. On the island of Samos, the Greeks built one of their earlier tunnels in 687 BC as a water conduit; it was discovered, in excellent condition, during excavations at the end of the last century.

The Romans, first-class engineers, developed conduit-tunnelling to a science parallel with that of constructing elevated aqueducts. Their masterpiece was a 3½-mile tunnel, ten feet high and six feet wide, hacked through Monte Salviano over 2,300 years ago to provide a run-off for Lake Fucino.

In antiquity artificial caves linked by passages and tunnels were hewn out of mountains and rocky ground for various purposes. Masada, the sheer-sided mountain which formed a natural fortress on the Dead Sea, had served as a hide-out and stronghold to many generations of desert tribes before Herod the Great, in the first century BC, built his palace near the summit and added huge underground granaries and cisterns. They enabled a thousand Jewish die-hards to hold out for three years against the Romans, after the destruction of the Temple by the legions of Titus in AD 70.

During the Middle Ages tunnelling and other branches of engineering stagnated. But the tunnelling art was revolutionised in the early 1600s when gunpowder was first used to blast and remove rock, a technique which rapidly replaced such ancient methods as lighting a fire at the rock face and then dousing the hot rock with water which caused the rock to crack and splinter.

The famous cartoon, anonymous and undated (probably 1804), showing the invasion of England by a fleet of warships, a squadron of troop-carrying balloons, and an army of infantry and artillery moving through a tunnel under the Channel . . .

The modern age of tunnelling may be said to have begun in 1707 with the construction of the Urner Loch, on the St. Gotthard road, Switzerland, which was blasted out of the rock to a length of 2,000 feet. Yet even gunpowder could not enable the Italian engineers to complete the series of tunnels through the maritime Alps between Genoa and Nice which had been started and abandoned in the fifteenth century. Work was resumed in 1795, but technical and political difficulties led to its abandonment after some 8,500 feet had been completed.

It was not until the turn of the nineteenth century, when the steam railway had yet to be invented, and engineering technology was still in its infancy, that the most imaginative tunnel scheme ever envisaged was born – the Channel Tunnel.

Modern technical and scientific history is a mere three centuries old, yet fables and misconceptions abound. The Channel Tunnel seems to have had a particular attraction for the myth-makers. One story was that a young French scientist by the name of Nicolas Desmarets started it all.

In 1750, Desmarets won the Academy of Amiens gold medal for the best essay which was published in 1753 as *'The Ancient Junction Between England and France, or the Straits of Calais; its formation by the rupture of an isthmus; its topography and geological structure'*. It was a remarkable work considering that its author could not have afforded to do much field research. However, nowhere does the essay suggest any artificial link between England and France – such as a tunnel.

The credit for having first thought of the Channel Tunnel was bestowed on Nicolas Desmarets some time after his death. His Amiens essay was reprinted in 1875 – but a preface had been added:

> At this moment, when the efforts of two great nations are being directed towards the realisation of that admirable project, the direct railway link between France and England, a book written over a century ago and since completely forgotten and which is intimately connected with that subject, cannot fail to arouse lively interest.
>
> It is, above all, clearly a work of historical and physical examination, for 125 years ago no one could have dreamt of re-establishing a junction destroyed by circumstances of chance . . .

The title page of the booklet reveals that the publication was sponsored by Messrs. McKean & Co., 'The inventors and owners of McKean's Rock Drill', whose Paris address (there was also one in the City of London) happened to be the same as that of the publisher, M. Isidore Liseux.

First aerial crossing of the English Channel. M. Blanchard and Dr. J. Jefferies departing from Dover,
January 1785.

It was a stroke of inspired publicity on the part of Messrs. McKean & Co. which led to Desmarets' posthumous fame. They provided the reprint of his essay with a wrapper saying in bold letters, *'Tunnel de la Manche'*, and contemporary journalists must have taken it for granted that the booklet contained the earliest suggestion of the scheme. They did not take the trouble to read the preface, or Desmarets' text. Hence the legend.

Then there is that famous cartoon, unfortunately anonymous and undated, showing an invasion of England by a fleet of warships, a squadron of troop-carrying balloons, and an army of infantry and artillery moving through a tunnel under the Channel. The usual date to this cartoon is 'around 1800', but many captions are more specific, saying, for example, 'A fantastic plan for the conquest of England, by the Frenchman J Ch. Thilorier'. Jean-Charles Thilorier was a well-known lawyer, born in 1750, who made his name in 1786 by successfully defending Cagliostro in the trial which ended the affair of Queen Marie Antoinette's necklace.

Thilorier was also an amateur scientist and inventor, and one of his schemes was an adaptation of the Montgolfiers' hot-air balloon. Writing in the Paris journal *Moniteur* in 1797, he offered to build 'a [fleet of] Montgolfière [balloons] large enough to carry into the heart of England an army which would be able to conquer her', and all for the trifling sum of 300,000 francs. The scheme met with general ridicule, but he returned to the challenge in 1803, suggesting that it should be financed by popular subscription. He called his invention the 'Thilorière', and a contemporary picture of it can be found in the *Bibliothèque Nationale*, Paris. This collection also contains the earliest known reproduction of that triple-invasion cartoon, with the descriptive caption:

> Several projects for the invasion of England, presented by some more or less ingenious inventors, with the object of carrying an expeditionary force to England. Among them are the Tunnel under the Channel, transport by *aerostats*, etc. Anonymous etching, probably 1804.

Alas, nothing to do with the Thilorière; for a closer look at the balloons on the print reveals that they are not open-ended hot-air balloons *à la* Montgolfière but closed 'Charlières', filled with hydrogen – similar to the type in which M. Blanchard and Dr. J. Jefferies made the first crossing of the Channel by air from Dover Castle to France on 7th January, 1785. Thilorière did not think of the Tunnel; but by 1802 someone else had.

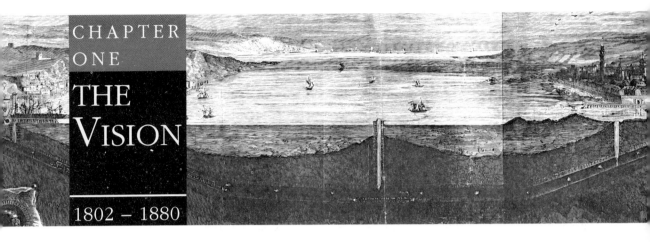

THE FIRST PROPOSAL

Early in 1802, the Peace of Amiens seemed to have ended the armed conflict between France and Britain for a long time to come. Many Englishmen took the opportunity to visit France again to see what it was like under its First Consul Bonaparte; among them was Charles James Fox, the distinguished Liberal statesman and leading advocate of what was much later called the *entente cordiale*. His main reason for going to Paris was to do some historical research for a work on the Revolution, but the French saw his visit as a gesture of friendship, and as his secretary, Mr. A. Trotter, recalls, he was fêted wherever he went.

Napoleon liked Fox, indeed as he reminisced in St. Helena: 'I loved him. He was the model of a statesman, and sooner or later his school ought to rule the world . . . we talked often, and without prejudice, about all kinds of things.' One of them was the Channel Tunnel.

The First Consul, relates Trotter in his memoirs, had reviewed plans for a Channel Tunnel submitted by an engineer, and was 'intrigued by the possibilities of the project'. Fox responded enthusiastically, exclaiming: 'This is one of the great things that we will be able to do together!' The French account of this conversation, not surprisingly, claims that it was Napoleon who said: '*C'est une des grandes choses que nous pourrions faire ensemble!*'

But it was not to be, for the Peace of Amiens lasted barely a year. War once again erupted, and in 1815 Waterloo saw the defeat of Napoleon.

The history of the Channel Tunnel has always been a mirror of European history, and particularly of the changing relations between France and Britain. Napoleon might well have had other ideas in mind when he discussed the Tunnel scheme so eagerly with Fox for, two years after hostilities had been resumed, preparations for invasion were observed along the French coast. But Napoleon abandoned any such thoughts after 1805. The defeat of the French Fleet at Trafalgar no doubt played a part in this decision, for to have contemplated invasion without naval supremacy would have been sheer folly. However, fear of invasion, irrational or not, was to haunt British military thinking well into the twentieth century.

Strangely enough, we know little about the man who submitted those plans to Napoleon in 1802, except that his name was Albert Mathieu[1] and that he was a mining engineer in the North of France. There is no record of his birth, or of what became of him subsequently. What we do know, however, is the kind of Tunnel he proposed; and for this we have to thank Thomé de Gamond, the great nineteenth century champion of the Tunnel. In de Gamond's major work on the subject, published in 1857, he wrote that to his knowledge Mathieu had conceived the plan 'at the end of the eighteenth century', and that sectional drawings of it had been exhibited first at the *Palais du Luxembourg* in Paris, then at the School of Mining, and eventually for a number of years at the *Institut*. Later, however, the drawings were withdrawn, by Mathieu himself or by his family, and, writes de Gamond, 'attempts at re-discovery in various archives were fruitless'.

Fortunately, de Gamond had been a pupil of an eminent professor of geology, Louis Cordier, who, after nearly fifty years, still remembered every detail of Mathieu's plans, and told his former student about them. Cordier had seen them when they were being exhibited at the *Institut*, where he had worked as a young teacher. Using traditional mining techniques, Mathieu had proposed a paved, vaulted underground passage for horse-drawn transport, lit by oil lamps and ventilated by a number of iron 'chimneys' protruding above the waves. A second, narrower tunnel would run below the paved passage for the purpose of drainage, emptying its water into reservoirs at either end. The two parallel tunnels would rise at a slight angle from each coast to the middle of the Straits, on to the Varne Bank, which Mathieu wanted to transform into a large artificial island, with an 'international town' and harbour for ships of all nations. The construction of the Tunnel, he thought, would be accelerated if tunnelling were to proceed not only from the coasts but also from the island on the Varne Bank: apart from the construction advantages, the centre island would provide the 'very necessary relay stations to change horses'.

Napoleon Bonaparte.

---

[1]Also known as Albert Mathieu-Favier.

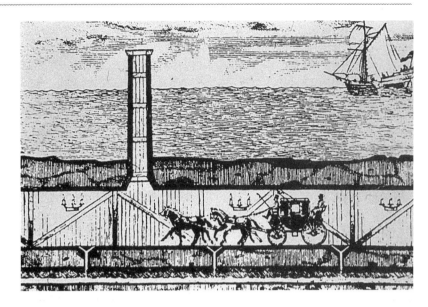

Albert Mathieu's scheme.

Mathieu certainly had ideas ahead of his time, although, in Cordier's judgement, he had not based his scheme on serious geological studies. Nevertheless it was the beginning.

THE ADVENT OF THE RAILWAYS

In 1820 the patent for an improved method of rolling wrought iron rails was granted to John Birkinshaw of Durham. Even though the railway had been used for half a century in colliery and quarry workings, the improved rails represented a major breakthrough, and were used by George Stephenson on the Stockton & Darlington and Whitstable & Canterbury lines. The former carried the first passengers on 10th October, 1825, and the train was drawn by a steam engine[2]. However, it was the opening of the Liverpool & Manchester railway in 1830 that first impressed the nation that a revolution in methods of travel had really taken place. It was for this line that Stephenson's *Rocket* was adopted, drawing a train weighing 13 tons over a distance of 35 miles at a speed of nearly 44 m.p.h. Eight further such engines were built by the Stephensons during the same year.

In 1830 the Duke of Wellington, then Prime Minister, expressed jaundiced views about the new railways, which would, he said, 'enable the working classes to move about unnecessarily'. But his concern went beyond mere social comment. He foresaw that steam propulsion was readily adaptable to ships and he

---

[2] 1828 saw the first railway line opened in France.

was particularly opposed to the building of a railway between Portsmouth and London, as he had the nightmare vision of the French fleet steaming into Portsmouth harbour and the disembarked army catching the 8.15 to London to take the capital by storm!

Despite his fears – and he was not alone in them – one can be certain that the advent of the railways provided a major incentive to early Tunnel visionaries[3], and in particular to Thomé de Gamond, the greatest of them all.

Thomé de Gamond.

THE PIONEER

Five years after Napoleon had seen Albert Mathieu's plans, Joseph-Aimé Thomé (later de Gamond) was born to an educated and well-to-do family in Poitiers. Thomé studied a variety of subjects at the universities of Prague, Vienna, and Ausburg. He was a brilliant student, obtaining a doctorate in both medicine and law and also becoming *officier* of military engineering. He returned to France to complete his studies in geology and civil engineering. It was then that he met and became friends with Prince Louis Bonaparte, later Napoleon III. Shortly thereafter he travelled to Egypt where he explored the possibility of 'piercing the isthmus' – later to become the Suez Canal[4].

In 1831 Thomé married the daughter of a distinguished French lawyer, de Gamond, and joined her

---

[3] The first to propose a submarine railway was a Glaswegian by the name of Rettie, in 1838, although it was not until the 1840s that the Englishman de la Haye put forward a scheme for such a railway. The scheme would cost £8 million, including the building of coastal railway stations 'on a magnificent scale'.

[4] A waterway of this kind appears to have been built during the reign of King Seti I, circa 1300 BC, which Napoleon, during the French expedition of 1798, had planned to reconstruct; but Nelson's victory at the Battle of the Nile resulted in his departure from Egypt. In many ways the history of the Suez Canal resembles that of the Channel Tunnel with its enthusiasms fluctuating throughout much of the nineteenth century. However, there is a fundamental difference between them, the Suez Canal being a passage for ships from one sea to the other, and the Channel Tunnel being a link between two land masses.

maiden name to his. He was now a man of independent means, free to concentrate on geological and hydrological problems. One of his pet ideas was the utilisation of the rivers of France for industrial purposes.

> Our investigations [he recalled] extended by degrees to the shores of the ocean . . . It was during the geological investigation of the Channel coast that we were first struck with the idea of a means of communication between England and the Continent. From the aspect of each shore, this communication appeared to us to have formerly existed naturally . . . was it not then possible for human industry to re-establish between the two lands an equivalent for this ancient connection?

It was this 'communication between England and the Continent' which fascinated de Gamond more than any other scheme, and in 1833 he began the technical and geological studies which were to occupy him for over forty years.

That same year, at the age of 26, he undertook soundings by hand line between Dover and Calais, recording the immersed lengths. Previous works by W. Phillips (1818) and Henri de la Beche (1821) indicated that the coastal strata of Britain and France were identical. This posed the question: was the Straits of Dover the result of the sinking of a rift valley, or the result of the erosion of an isthmus which had previously connected Kent with Northern France?

As early as 1834 de Gamond had his first scheme ready. This was the laying of iron tubes lined with brick in sections across the Straits, but it was abandoned because of the cost of levelling the sea bed. Further schemes included a Channel bridge, using forged iron and cast iron (steel was not yet available in large quantities); a floating bridge; and a bridge resting on granite piles with arches of syenite; but again, the cost was prohibitive. He also suggested 'narrowing' the Channel by two jetties, each eight kilometres long, using 'an enormous steam-powered boat' to ferry passengers and freight between the two jetties. In 1840 de Gamond embarked on yet another form of bridge scheme, this time an artificial isthmus, leaving open three wide channels for shipping. However, his chief encouragement towards a tunnel came from Sir Marc Isambard Brunel's work on the tunnel under the Thames, between Rotherhithe and Wapping, despite all the setbacks suffered by that enterprise. It was begun in 1825; the river broke through twice, the first time in 1827, the second in 1828, and after this, work was not resumed until 1835. Part of the ground penetrated was almost liquid mud, and the tunnel was not completed until 1843.

It would do less than justice to the history of the project to ignore the other nineteenth-century engineers on both sides of the Channel who proposed cross-Channel links, and an account of these will be found in Appendix I. But whatever the nature of the proposal, de Gamond had either thought of it first, or examined it later with the unprejudiced mind of a scientist.

The Great Exhibition of 1851 softened England's xenophobia, at least for a while. With the peoples of the world flocking to London as visitors and exhibitors, to be outward-looking became fashionable. Later that year, however, events took a dramatic turn when Prince Louis Bonaparte, whom de Gamond had known so well since their student days, assumed near-dictatorial powers by a *coup d'état* and took the title of Emperor Napoleon III in 1852.

Artist's impression of 19th century plans for a Channel Tunnel.

A bizarre scheme for a Channel Tunnel proposed by Hector Horeau in 1851.

In 1855, to obtain his own geological evidence, de Gamond made a series of dives to bring up samples near Boulogne and Folkestone, as well as from the Varne Bank. Naked save for ten pigs' bladders for buoyancy and 160 pounds of stones for ballast tied around his waist, and having filled his mouth with olive oil so that he could expel air without water being forced into his lungs, he reached depths of nearly 100 feet. His account shows a great capacity for scientific observation:

> In these rapid descents I could only cast a furtive glance over the bed of the sea, which in this place was quite dark when the sun was obscured. This darkness arose from the very deep colour of the ground in that region. But when the sun shone and this was the case during my first two descents – the liquid medium assumed a rather milky appearance, transparent enough, and one could very well distinguish the remains of white shells, with which the dark bed of the sea seemed to be strewn. I even saw spotted bodies pass with a rapid movement, and these I judged to be shoals of flat fish of the sole or skate family, disturbed by my presence.
>
> It was on ascending from my third and last visit to the bed of the sea that I was attacked by some carnivorous fish, which seized me by the legs and arms. One of them bit me on the chin, and would at the same time have attacked my throat, if it

had not been preserved by a handkerchief. I ridded myself promptly of this one, which caused me sharp pain, and which left me as soon as my hand touched it. I thought myself lost. However, preserved more by instinctive energy than by an act of volition, I was fortunate enough not to open my mouth, and I reappeared on the top of the water, after being immersed fifty two seconds. My men saw one of the monsters[5] which had assailed me until I had reached the surface.

The following year de Gamond submitted his scheme to Napoleon, who passed it for consideration to his Scientific Research Commission. After asking for the Commission's blessing on his study, he said, modestly: 'We have advanced up to the limits of our personal capacity. Now it is necessary to have this work continued by a collective of intellects.'

De Gamond later expressed his satisfaction with the Commission's reactions: 'The Commission looked upon the study from the practical angle, coming to the conclusion that the material realisation of the project required the co-operation of the Government.'

It is characteristic of the man that he described his achievement as a mere 'study'. In fact, his *Étude pour l'avant-project d'un Tunnel sous-marin entre l'Angleterre et la France*, published in 1857, is a work of vision and genius. Apart from being scientifically sound, it examined the problem from every point of view – geological, technical, economic, financial and administrative. The opportunity to deride the 'crowd of raw projects' offered by 'pirates' less qualified than himself had been a powerful motive for de Gamond in publishing his work.

Perhaps the most fascinating part of the study is the very beginning; remembering that railways were only a quarter of a century old at the time of writing:

> The construction of a road link between England and France, an idea conceived before the end of the last century, has become the object of renewed interest, especially since the introduction of the railways gave such great impetus to transport in these two countries.
>
> If one takes a look at the map of those new roads, abruptly intercepted by the sea, one grows convinced that their heads, at both sides of the Straits, are no more than half posts destined to be joined in a common and continuous transport system . . .
>
> The creation of such a route is not an isolated concept; it is the complementary link of a great current of traffic among the nations, a current which extends across Europe in parallel branches, converging on the Mediterranean and then turning towards the Orient to penetrate into India, thus extending

---

[5] The 'monsters' are believed to have been conger eels.

towards the two poles of England's possessions. There are, however, three natural obstacles which seem to cut across that grand highway of the nations.

1)   The deserts of Lower Asia . . .
2)   The wall of the Alps – the roads of France and Italy will be linked up by a subterranean gallery under the ridge of Mont Cenis[6];
3)   The Straits of Dover. This is the object of the scheme we propose; what it aims at is the joining of the English island to the European Continent by a permanent road dug under the sea.

Thomé de Gamond had investigated all the possible forms of a link. In his submission he discussed the two most feasible alternatives before offering his own choice, that of a bored tunnel[7]. However, his tunnel – from Dover/Folkestone to Cap Gris Nez – like Mathieu's included an intermediate 'maritime station' on the Varne Bank. The *Étoile de Varne*, as he called it, would have a lighthouse, an inner port of refuge of seven hectares, ten hectares of outer quays, a telegraph office, and other amenities.

Thomé de Gamond's preference for a bored tunnel with an intermediate maritime station on the Varne Bank.

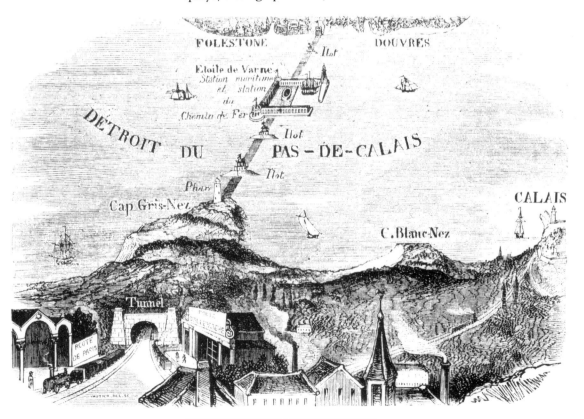

---

[6] The 5-mile-long Mont Cenis tunnel was begun in 1857 and opened in 1871.
[7] The original 'boring' machine was patented in 1875 by Colonel Frederick Edward Blackett Beaumont, MP, RE, although boring methods had existed for some time.

He reported the results of his 1,602 geological samplings and made reference to Brunel's Thames tunnel, which had yielded so much experience of boring techniques and machinery. Construction would be carried out in nine sections from 12 temporary islands built of stone and would take six years. The Tunnel would be stone-vaulted, measuring internally seven metres high and nine metres wide. The lower part would carry a double railway track, with narrow footways on either side. A drainage duct would run underneath the tracks.

Thomé de Gamond took into consideration half a dozen routes for the Tunnel, before deciding on the 33-kilometre (21-mile) route from Cap Gris Nez to East Wear Bay, crossing the northern tip of the Varne Bank. He paid special attention to the approach lines, linking up with the Dover-to-London railway and on the French side with the Boulogne-Amiens-Paris line. Three ventilation shafts, one at either end of the Tunnel and the one at the *Étoile de Varne*, were considered sufficient. The Tunnel would be gas-lit from gasometers near both entrances.

The most modern features of de Gamond's study were his economic estimates and traffic forecasts. The total expenditure, to be spread over six years and borne by the Governments of France and England in equal parts, was estimated at 170 million gold francs (then equivalent to about £8½ million), at 6% interest. Passengers would pay 12 francs per head, excess baggage would cost 100 francs per ton, and freight carried by slow trains 12 francs per ton.

His traffic forecasts were based on the sea crossings by steam ships carrying some 350,000 passengers a year. He noted that in the early 1840s out of every hundred passengers 76 were English and only 24 Continental, suggesting that the seafaring race was more prepared to put up with the inconvenience of a sea crossing; so the Tunnel would draw new customers from a large potential market on the Continent. He estimated that 800,000 travellers would use the Tunnel every year, and the daily figure would run to 2,192 passengers, conveyed by five ten-coach trains in each direction. The annual income including freight charges would amount to 26.2 million francs (£1.3 million) – and show a handsome profit margin above operating and maintenance costs, capital repayment and interest.

Copies of the *Étude* were sent to Government departments, statesmen, leading engineers, scientific and technical associations on both sides of the Channel. One copy is still to be found in the Library of the Institution of Civil Engineers in London with an accompanying letter in Thomé de Gamond's hand:

In daring to place under your auspices this humble work, I hope to obtain your indulgence for all that is imperfect in it. It is but a preparatory study before obtaining more precise information, a work which can only be continued from this point with the will and united resources of the two powerful nations. I have reason to think that we shall soon see an international effort to this object . . . an end as useful as it is glorious.

Reactions in both countries were favourable. 'Well done!' wrote Professor Cordier to his 'dear disciple', who had well remembered Mathieu's scheme. The French press waxed enthusiastic. 'England to become continental – what a beautiful dream!' exclaimed the *Siècle*. 'England united to the Continent by a Submarine Tunnel, which will join without interruption all the French, German, Belgian, Russian and other railways with those of Great Britain! The sea crossed in twenty-five minutes by rapid and incessant trains . . . Well, this dream may become, in five or six years, a reality.' The *Ingénieur* declared that de Gamond had 'presented to the elevated feelings of the age a project worthy to satisfy them', though he had failed to solve the problem of 'submarine invasion'. Those in authority were friendly but non-committal. But what was 'most precious' to de Gamond was the sympathy of men like the far-sighted economist, mining engineer and senator, Michel Chevalier, who was later to play an important role in the Tunnel project.

The English press gave the work a full and sympathetic coverage. A typical comment was that of the *Illustrated London News*, which emphasised the economic aspects:

Early tunnelling methods!
Marc Isambard Brunel's
Thames tunnel shield.

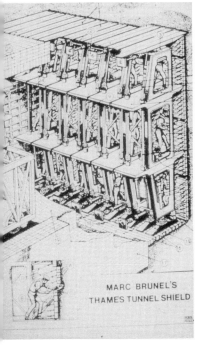

MARC BRUNEL'S
THAMES TUNNEL SHIELD

> The creation of this great line of junction would put an end to the commercial isolation with which England will be threatened by the completion of the great railway system which will connect without breaking bulk the centre of Europe with the western ports of the Continent . . . [It] will permit England to sustain with advantage the competition which will be created by the junction of the ports of the west of the continent with the centre of Europe, by means of the railways which are in course of execution.

De Gamond's moment of triumph was short-lived. A dramatic event, which occurred in 1858, soured relations between England and France for nearly a decade. De Gamond describes this affair in his *Account of the plans for a new project of a Submarine Tunnel between England and France*[8], originally written for the *Exposition Universelle* of 1867 in Paris. His plans had been well known amongst British engineers for many years and he recalled:

> It had appeared to us opportune first to obtain the effective co-operation of the principal English engineers. This was already

---

[8] A later edition was published in London in 1870.

done so far as Isambard Kingdom Brunel and Joseph Locke were concerned. We had known the first from our youth, and had received from his father the kindest support in 1827, at the time when he was making the Thames Tunnel. Besides, we had made with his young brother engineer a visit to the Caledonian Canal from one sea to the other, and had been present at the opening of the Liverpool and Manchester Railway. We had long since secured the aid of Joseph Locke[9]. We had known him first when he was engineer of the railway from London to Southampton . . . Locke was able to bring a considerable contingent of English elements into the formation of that company.

As to Robert Stephenson, he was already connected with French engineers when the first promoters of the Suez Canal . . . had shown to the world the possibility and opportuneness of excavating anew the ancient canal between the two seas . . .

The first time we had the opportunity of talking to Lord Palmerston on the subject of the Submarine Tunnel, we found him at first rather close: What! you pretend to ask us to contribute to a work the object of which is to shorten a distance which we find already too short! We expressed to him our wish to talk of it to Prince Albert in his presence, and to this he very kindly consented.

The Prince Consort had supported this project with truly enthusiastic sympathy. His reception was therefore most kind. He entered into conversation, in which the Prince unfolded all the advantages which his elevated mind foresaw for England in the creation of a road to the Continent. Lord Palmerston, without losing that perfectly courteous tone which was habitual with him, made, however, a remark to the Prince which was very rude at bottom: You would think quite differently if you had been born in this island!

We were ourselves perfectly stupefied with his unexpected apostrophe. To make Prince Albert, whose love of the country of his adoption was well known, feel that he was a foreigner, was shocking to us, and we felt deeply hurt.

Some days after we went to excuse ourselves with the Prince Consort for having been the cause of this disgraceful incident. The Prince appeared not to have been offended, and told us that he had received this innocent dart as one of the frequent sallies in which *Pam* dealt. Then he added that he had said a few words about the Submarine Tunnel to the Queen, and Her Majesty had been graciously pleased to answer him in these good words: 'You may tell the French engineer that if he can accomplish it, I will give him my blessing in my own name, and in the names of all the ladies in England.'

Lord Palmerston knew this, and whether it worked a

---

[9] Born in the Sheffield mining district in 1805, Locke worked for Robert Stephenson's father, George, from 1823; he was in charge of the eastern section of the Manchester-Liverpool railway, with the Edge Hill tunnel. Locke was on the footplate of the *Rocket* at the opening of the line in 1830 when William Huskisson, MP for Liverpool, was knocked down and fatally injured. It was the first railway accident.

sudden change in his thoughts, considering it as public opinion, or whether he was not really so disposed to resistance as he had appeared to be, he said frankly one evening at a numerous party, at which the Tunnel was spoken about: 'This project will be carried out because it is respectable, and because it is favoured by all the ladies of England.'

Napoleon III

On his side, Richard Cobden had said at a public meeting: 'I consider the submarine Tunnel as the true arch of alliance between the two countries.' The question was making its way, therefore, in the public mind of England when there came to pass in Paris, in the vestibule of the Opera, on the 14th January, 1858, an odious attempt which adjourned it for a long time. Some revelations made public after this sad event threw some clouds over the Governments of the two countries. The Ambassador of France, Count de Persigny, who had begun to aid us in London, was recalled, and replaced by a soldier.

We then received from a high personage, to whom we are attached by the precious recollections of youth, dating from exile, the advice to avoid using up the question of the Submarine Tunnel by persisting in introducing it under such inopportune circumstances. He had even the kindness to add that it would be equally to be regretted if we used ourselves up uselessly in pursuing an idea which we ought to know how to keep for better times.

The 'odious attempt' to which de Gamond referred was an attempt made by Orsini – an Italian – to assassinate Napoleon III. But the would-be assassin had come from England as had his bomb. Orsini had hoped that the death of the Emperor would start a revolution in France which would assist Italy in its struggle for freedom and unification. Napoleon, aware that he was widely unpopular, introduced a spate of new emergency laws: foreigners were deported, civil liberties curtailed. Understandably, relations with Britain were strained. The 'high personage' who advised de Gamond to lie low for the time being may well have been his old friend Napoleon himself.

In April 1866, the *Moniteur Universel* reported that a 'campaign had begun on the Straits of Dover for the last verifications of a project for a Submarine Tunnel . . . ' and that the plans would be shown at the Paris Universal Exhibition the following year. Thus the Tunnel, which had remained 'discreetly in the background' for nearly a decade, reappeared in a new, and amended form.

De Gamond had been well aware of the criticism which his 1857 project had aroused, and of two 'grave objections' in particular. First, the 'rather puerile' fear of the English that the French would invade through the Tunnel – a fear which he had sought to allay by incorporating valves for flooding the Tunnel 'by the sudden introduction of the sea'. Second, the objection 'that the construction of a tunnel of this length would last for ages, and that we should never see it finished.' So he omitted the valves and the phased construction from islands from his amended scheme for the *Exposition Universelle* of 1867. 'We confess', he wrote, 'that we introduced the islands without thinking that it would ever be necessary to apply them' and that the Tunnel would be best driven in four sections: in both directions from a 'maritime workshop' on the Varne Bank and, of course, from both shores. As to the valves, he believed – mistakenly, as it turned out – that 'no one now fears the furtive invasion of an army'.

Thomé de Gamond also revised his traffic estimates upwards: within a dozen years of the opening of the Tunnel, one million passengers would be carried annually. 'It cannot be doubted that the great facility of travelling in the same train between Paris and London by the Tunnel would cause a new tide of traffic to set in, the amount of which is incalculable.' Excess baggage would amount to something like 8,000 tons, while three million tons of freight could be regarded as a conservative estimate. Construction costs on the other hand would be reduced to £8 million because open-cutting approaches would replace the approach tunnels proposed earlier. In the first year of operation receipts would surpass £900,000 and continue to rise.

'From the preceding data', concluded de Gamond, 'it will be seen that the Submarine Tunnel is destined to become an extremely profitable undertaking, and that the necessary capital might be usefully applied to its execution.'

Copies of his proposals were distributed at the *Exposition Universelle*, and a large display attracted a constant flow of visitors. It would appear that he met William Low[10], a mining engineer and colliery owner in Wales, on this occasion, though it is not known whether de Gamond knew that Low had personally submitted his own plans for a Channel Tunnel to Napoleon III on 23rd April before the exhibition opened.

It was in the early 1860s that Low became interested in the Channel Tunnel and he soon identified de Gamond's problem of ventilation. His own plan, based on his experience as a mining engineer, included the principle of 'upcast' and 'downcast' shafts, which also provided alternative escape routes for the workmen in

---

[10] Born in 1814 at Rothesay, Isle of Bute, William Low spent his apprenticeship in Glasgow, and then worked as an assistant engineer on the construction of the Great Western Railway under Isambard Kingdom Brunel. Next he helped to prepare the plans for the London & North-Western Railway to Holyhead and other main lines. Later, he suggested the building of a bridge over (instead of a tunnel under) the River Severn, and his plans for a tunnel under the Mersey received Parliamentary sanction.

Artist's impression of the British engineer William Low discussing Tunnel matters with Napoleon III.

case of an accident. This meant the construction of twin tunnels, connected by a number of cross-passages, to ensure a continuous circulation of air.

Napoleon, according to a witness, listened carefully to Low's explanations: and remarked: 'I see perfectly – you will give us English air through one tunnel and we shall send you French air through the other!' The Emperor concluded the meeting on an optimistic note. He was prepared to consider the project: 'I should be glad to see it done. As soon as your prospectus is ready, send it to me, and I will lay it before my Minister and do all I can to support it.'

Low foresaw the necessity for Anglo/French co-operation and said in his published report:

> In an International work of this nature, it was obviously desirable to obtain the co-operation of a French engineer, and it was quite as obvious that M. Thomé de Gamond was by his long study of the question the person to whom to apply for co-operation. Mr. Low therefore put himself in communication with M. de Gamond, who immediately, without the slightest reservation, put his geological studies and sections at Mr. Low's disposal.

THE ANGLO/FRENCH COMMITTEE

The co-operation between the two engineers quickly expanded into an official full-scale Anglo/French venture. Low got in touch with two more British engineers whom he knew to be interested in the Tunnel scheme, James Brunlees[11] and John Hawkshaw[12] (both later knighted). The two engineers co-operated on many hundreds of miles of railway in Scotland and England. They frequently differed on technical and personal grounds though never as fiercely as Low and Hawkshaw in later years.

Brunlees, Hawkshaw and Low signed the report of what had originally been Low's individual proposal to Napoleon, who granted another audience to the Anglo/French delegation – some twenty strong – in the Summer of 1868. It represented the newly formed 'Anglo/French Channel Tunnel Committee', including its chairman, Lord Richard Grosvenor, its secretary William Bellingham, the engineers William Low, Thomé de Gamond, James Brunlees, Paulin Talabot, John Hawkshaw and Senator Michel Chevalier. The

---

[11] Brunlees, born at Kelso in 1816, began his career as a road surveyor. In 1838 he worked on the survey for the Bolton-Preston railway, and on the tracks being built towards Glasgow and Edinburgh. In 1865 he helped to build the Mont Cenis Summit Railway, and, in the 1880s, the Liverpool-Birkenhead tunnel under the Mersey.
[12] Born in Leeds 1811. He constructed the Severn Tunnel, Holyhead Harbour, and the Charing Cross and Cannon Street railway stations and bridges in London. He died in 1891, but had lived long enough to play a role in the first part of London's underground railway system.

Committee commanded powerful British support, headed by the Archbishop of York, the Dukes of Argyll and Sutherland, and some ninety men prominent in all walks of public life, including several Peers, a number of MPs, bankers, industrialists, directors of railways, mayors and judges. They had all lent their names to the 'Engineers' Report on a Proposed Tunnel', expressing 'the most earnest wish for the speedy success of this fruitful work', and hoping that the French Emperor, to whom the publication was addressed, would grant the scheme his 'august protection'.

At the request of the Emperor the members of the French group of the Committee, led by Michel Chevalier and the eminent engineer, Paulin Talabot, formed an official Commission to investigate the new Tunnel plans.

THE GOLDEN AGE OF THE TUNNEL

The cohesion of eminent like-minds, on both sides of the Channel, gave substance to the renaissance of interest in the great venture. The twin tunnels suggested by Low were technically an improvement and seemed to have solved the ventilation problem which made the Varne Bank 'maritime station' superfluous. The British engineers paid tribute to de Gamond's studies in their 1868 Report:

> Acknowledging the scientific accuracy of M. de Gamond's conclusions, the Commission were of the opinion that the investigations should be practically tested by sinking pits on the two shores, and driving a few short headings under the sea.

Artist's impression of a 19th century bridge scheme.

Hawkshaw had already carried out his own survey with the help of Hartsink Day, the eminent geologist. In 1865, he had an artesian well sunk to a depth of about 1,000 feet at Calais, and borings carried out near the South Foreland and at a point three miles west of Calais, and collected some 500 samples from the sea bed by means of tallow-coated leads. It seemed that the Upper Cretaceous bed was continuous from coast to coast. Hawkshaw's report concluded that on the proposed Tunnel line there was probably no great fault in the continuity or regularity of the strata between the two shores, and he recommended that the Tunnel should be driven through the Lower Chalk stratum.

> There seems to be no reason to assume [concluded the British Engineers' report] that the Tunnel would cost more than ten million sterling, or that it could not be completed in nine or ten years.

In March 1869 the British and French Governments declined requests made by the Anglo/French Channel Tunnel Committee for a guarantee of interest on the capital required for exploratory works, or any other financial obligations. A year later, the Committee, undaunted, applied to the French Government for the sole concession to build and operate a Submarine Tunnel, so as to secure some official authorisation when seeking the necessary private capital. The French Ambassador in London, the Marquis de la Valette, was instructed by his Government to sound out the British Foreign Secretary, Lord Clarendon, who subsequently proved non-committal; he wanted to know more about the cost and viability of the project, and the attitude of the French Government. Both sides temporised: Paris wanted a more definite statement from London. Clarendon referred back to the Board of Trade, who replied on 14th July, 1870, that it had nothing to add to its previous statement. All of which was of little consequence as, a few days later, war broke out between France and Prussia.

The Franco/Prussian War, the capture of Napoleon III and the proclamation of the Prussian King as German Emperor at Versailles, obscured any further thoughts of the Tunnel, though not for long. Public opinion in England, initially sympathetic towards the Germans, swung in favour of the French, particularly the Parisians, who had suffered dreadfully during the uprising of the Communards. Now that Napoleon had been deposed, the Third Republic was considered a neighbour whose friendship was needed to counterbalance the new and powerful united Germany.

Thus, it seemed timely that the Anglo/French Channel Tunnel Committee submit a new application for the Tunnel concession to the French Government in the Summer of 1871, complete with cost estimates worked out by Low, Hawkshaw, and Brunlees. Significantly, Thomé de Gamond's name was missing. There had been some behind-the-scenes disagreements between the engineers. Hawkshaw favoured a single tunnel, while Low would not be moved from his twin-tunnel concept. De Gamond, who had first supported Hawkshaw, had switched his allegiance. In his first letter to Low in August 1867 he wrote:

> I see no serious obstacle to the construction of a Tunnel on the line you propose, and I have . . . no exclusive preference for one project or another. My preference in the course of this study has often varied in favour of this or that method of operation, as it must vary in the mind of any conscientious engineer who seeks light . . . I was the solitary student of this enterprise . . . I have seen therefore with the liveliest satisfaction that this question has attracted the attention of British Engineers during the last few years . . . nothing can be done in France while the English do not take the initiative. But I have always hoped to live long enough to become the coadjutor of an English engineer who would understand the great utility of this work . . . Providence has willed that you

should be that man, and you have on your side the happy thought that the association with a French engineer grown old in the study of this question might be useful in the execution of it.

However, the British engineers were clearly no longer interested in the 'mental aid' of their old colleague, and he faded from the Tunnel scene – apparently not too greatly disappointed and happy that the scheme now seemed to be proceeding along more positive lines. Having spent all his private money on the Tunnel scheme, he lingered in poverty, supported only by his daughter's income as a teacher of piano. He died in 1876, four years before the first attempt to construct a Channel Tunnel was made.

The next step for the Anglo/French Channel Tunnel Committee was to form a company to proceed with the venture. Without any encouragement from either Government, the 'Channel Tunnel Company Limited' was incorporated and registered in London on 15th January, 1872, with the objective: 'To construct an underground tunnel beneath the Straits of Dover, between England and France', including a railway and an electric telegraph line[13]. The Company would also 'maintain and work' the railway and telegraph through the Tunnel, acquire land in England and France and obtain from the authorities in the two countries any powers or privileges necessary for the purpose. The capital of the Company was £30,000, divided into 1,500 shares of £20.00 each.

There was no parallel Company in France as yet, and the application to the French Government for the sole concession to build and operate the Tunnel was still pending. But surprisingly and suddenly all the pieces fell into place.

The 1870s were a strange decade in the history of the Tunnel. Even though both Governments were theoretically in favour of the scheme, their ideas on how to proceed were noticeably different. The situation was further exacerbated as a result of a wrongly addressed letter. When the Channel Tunnel Company – whose status was still nominal – applied to the Board of Trade for permission to drive an experimental heading under the foreshore east of Dover, the Board demanded detailed plans. This written demand was wrongly addressed to Hawkshaw's office instead of to the Company, whereupon Hawkshaw drew up a plan showing the Tunnel starting from St. Margaret's Bay – about a mile north of the site chosen by William Low. When subsequently asked to agree to this plan, Low was furious. Lord Grosvenor tried to reconcile the parties, but in vain. Hawkshaw refused to have anything more to do

---

[13] The world's first submarine telegraph cable had been laid under the Straits of Dover in 1851.

with Low who stormed out, taking with him his plan for a double-bore, self-ventilating Tunnel, leaving Hawkshaw and Brunlees as official engineers of the Channel Tunnel Company with its single-bore, double railway track plan. Low formed his own 'Anglo/French Submarine Railway Company' with the aim to banish its competitor from Tunnel business.

Meanwhile, the French Government set up a commission. Low attended its *enquête* at Arras in November 1873, and put forward his scheme, in opposition to the Hawkshaw/Brunlees plans which Lord Grosvenor had submitted as the official British project. The Commissioners, perhaps unsure of their brief, would admit Low only as a 'critic'. But he had the satisfaction of hearing the Commission's verdict that the plans submitted by the Grosvenor group were 'inferior in detail' to those he had presented in 1869, 'and but for the unprecedented nature of the enterprise they would not have been acceptable'. The Commissioners accepted that much had been done to establish the practicability of the work, but further research into the geology of the Straits would be necessary before a concession was granted, and concluded by recommending yet another Commission! By this time the French Government had an additional reason for stalling: the Tories, with Disraeli as Prime Minister, had returned to power in 1874, and it was necessary to make sure of the new Government's attitude.

Lord Derby, the Foreign Secretary, sought an opinion from the Board of Trade, who replied on 9th December, 1874:

> The Board of Trade can have no doubt of the utility of the work if successfully completed, and they think that it ought not to be opposed so long as the English Government is not asked to make any gift, loan or guarantee.

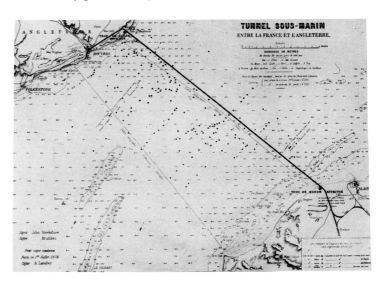

Tunnel route proposed by John Hawkshaw and William Brunlees in 1874. This tunnel was to extend between St. Margaret's Bay, in Kent, and a point near Sangatte.

The Board of Trade referred the military implications to the War Office, who replied succinctly that there were no objections to the undertaking on military grounds.

On 24th December, 1874, Lord Derby communicated the views of the British Government to the French Ambassador and also sent a formal despatch to Lord Lyons, Britain's Ambassador to France, requesting him to communicate to the French Government Her Majesty's Government's approval of the Channel Tunnel enterprise.

Never before had the subject of the Tunnel created such a flow of correspondence between the two Governments. Count de Jarnac, the French Ambassador in London, Lord Derby, the British Foreign Secretary, Lord Lyons, the British Ambassador in Paris, the Board of Trade and the Treasury, the French Ministries of Foreign Affairs and of Public Works, all exchanged views and forwarded documents with unprecedented intensity. Much of this correspondence, between October 1874 and August 1875, referred to the formation of a new Joint Commission. This Commission comprised, on the French side, Kleitz, Inspector General of Roads and Bridges; and two of his ministerial colleagues, Gavard and de Lapparent; and on the British side, Captain Tyler of the Board of Trade; Horace Watson, solicitor to the Department of Woods and Forests; and Charles Malcolm Kennedy of the Foreign Office. They began their sessions in London in March 1875, and their report was published in the Spring of 1876. On 30th May a protocol signed by both Governments accepted it 'as the basis of a Treaty to be concluded concerning the Submarine Tunnel'.

The Joint Commission's draft Treaty proposed regulations on jurisdiction during construction and after completion; the submarine frontier between Britain and France; the mode of exploitation; and the terms of an agreement to be concluded between the Channel Tunnel Company and its French counterpart, the *Société du Chemin de Fer Sous-marin entre la France et l'Angleterre* (formed on 1st February, 1875, with Michel Chevalier as its chairman). The French Tunnel Company was to be accorded a 99-year-concession by the National Assembly, to be effective five years after the granting of the concession. Despite the wish of the British Government to wait until the Joint Commission presented its report, the concession was granted on 2nd August, 1875, for the period of 1881 to 1980, under the condition that the British and French Tunnel Companies came to an agreement by 1881.

The French Tunnel Company had a capital of two million gold francs (£80,000), divided into 400 shares, half of which were subscribed by the *Chemin de Fer du Nord*, a quarter by the bankers de Rothschild Frères, and the remainder raised on the money market. The French Channel Tunnel Company requested no money from the State and no guarantee of interest, undertaking to carry out all the necessary works preparatory to construction. It would first contact its British counterpart in order to synchronise activities on both sides of the Channel.

RIVALRY IN BRITAIN

But what was the British counterpart? The Channel Tunnel Company? Or its rival the South-Eastern Railway Company? The latter had taken the place of Low's now defunct Anglo/French Submarine Railway Company and had received the Royal Assent on 16th July, 1874, enabling it 'to afford pecuniary aid towards the construction of borings and other works in connection with a

Tunnel under the English Channel'. By this time a new personality had begun to dominate the British scene, Edward Watkin[14], whose chairmanship of the South-Eastern Railway brought him into direct competition with the London, Chatham and Dover Railway. Both lines connected London with Dover; and Watkin's inevitable conflict with Lord Grosvenor, Chairman of the London, Chatham and Dover Railway, and his successor James S. Forbes, considerably harmed the Tunnel concept. Watkin accepted Low's scheme, while Lord Grosvenor stuck to the plans of Sir John Hawkshaw and Brunlees. In 1874 Parliament empowered the rival companies to spend £20,000 each – either together or separately – on Tunnel research. Watkin told Forbes that he was prepared to put up his share, but Forbes in his reply made all kinds of stipulations, particularly about the siting of the Tunnel terminal, and negotiations broke down. Thus the French Tunnel Company, which had in vain tried to reconcile the English railway rivals, remained in a quandary as to which would become its partner in the venture.

Sir Edward Watkin.

[14] Born in Salford in 1819. The son of a prosperous cotton merchant, Watkin joined his father's firm as soon as he had left school. Railways soon became his special interest, and in 1845 he became secretary of the Trent Valley Railway, where he negotiated the merger with the London North Western Railway. In 1853, he was appointed general manager of the Manchester, Sheffield, and Lincolnshire Railway, liaising with the Great Northern and the Midland Railways. He became Chairman of the South-Eastern Railway in 1866, a post he held for twenty-eight years, and also, from 1872 onwards, London's Metropolitan Railway. He was knighted in 1880.

On 2nd August, 1875, the Channel Tunnel Company Limited received the Royal Assent to its Bill. It could now acquire land and carry out preliminary works at St. Margaret's Bay. Unfortunately the influx of water was so great the works had to be abandoned. A tunnel could not be successfully driven from St Margaret's Bay. Clearly the Hawkshaw/Brunlees team had made a serious miscalculation in choosing the site to which William Low had so earnestly objected and the way was now open for Watkin and Low to advance their plans with all speed.

Her Majesty's Government may have looked on placidly at this turn of events, but Queen Victoria grew suddenly alarmed. Her abhorrence of the dreaded *mal de mer* was overtaken by her growing anxiety about a possible French invasion – regardless of the War Office's judgement that these fears were unjustified. She wrote to Disraeli in 1875, expressing the hope that 'the Government will do nothing to encourage the proposed Tunnel which we think very objectionable'. It was a very Royal *volte-face*.

The Channel Tunnel Company was facing further problems – financial ones. The backers were seeking more money to start digging at a new site. The 1875 Act stipulated that the acquisition of land for this purpose must take place within a year. Rothschild & Sons had already subscribed £20,000; the Channel Tunnel Company had put up the same sum; but a further £40,000 was required to match the French Company's expenditure. It could only be found in the money markets; but only £15,000 was raised before the concession expired. So it was the South-Eastern Railway, which appeared to have no financial problems, which became the British counterpart to the French Tunnel Company.

The French Company had commissioned a further geological survey from Messrs. Poitier and de Lapparent, together with the French Navy's hydrographer, Larousse. Employing the steam paddle vessel the *Ajax*, 7,600 soundings were made and 3,267 samples of the sea bed were obtained, enabling the experts in 1876 to draw a complete geological map of the Channel bed.

English geologists will be surprised at the extreme minuteness with which the French engineers have mapped out the bed of the Channel [wrote the eminent British geologist, William Topley, in his report on the survey], and some will possibly be inclined to doubt whether submarine mapping can be carried out with so much certainty. This doubt will arise most strongly with reference to the boundary lines of the sub-divisions of the Chalk. We must, however, remember that the greatest care has been taken in collecting specimens from the bottom . . . Nothing likely to help in arriving at correct results appears to have been left undone . . . I therefore think that, in future discussions, we shall be safe in assuming the general correctness of the map.

The paddle-steamer Ajax *taking samples from the sea bed in 1876.*

In Topley's opinion, the main points emerging from the French survey were: no tunnel could be bored in a completely straight line through the Lower Chalk; it must either pass through a one- or two-mile stretch of gault (clay and carbonate of lime), or through a stretch of chalk with flints near the middle of the Channel. The latter course seemed preferable, though it would mean a slight change of direction. The route advocated by Hawkshaw and Brunlees, St. Margaret's Bay to a point near Sangatte, still appeared to be 'the best which can be proposed', concluded Topley. But at the time he was writing, the St. Margaret's Bay site had been inundated and abandoned.

There were no such complications at Sangatte. By the Autumn of 1877, several shafts had been sunk to depths of up to 330 feet. The Lower Chalk had been reached. Water pumps and ventilators had been installed, and a few short headings were being driven, with the ease that the geologists and engineers had expected.

In France public confidence in the Tunnel scheme had risen to unprecedented heights. Founders' shares in the French Tunnel Company, nominal value 5,000 francs, were changing hands at 130,000 francs.

Preliminary works at St. Margaret's Bay, 1870s.

CHAPTER
TWO

THE
FIRST
ATTEMPT

1880 – 1882

In 1880, under Watkin's direction, the first shaft was sunk at Abbots Cliff, halfway between Dover and Folkestone, and a horizontal pilot gallery was driven through the cliff ten feet above high water. The plan was to enlarge this 'No. 1 heading' to the width of a railway tunnel connecting the South-Eastern line with the Tunnel starting further north at Shakespeare Cliff.

The wall of the shaft bears to this day a touching inscription crudely cut in the chalk:

THIS

TUNNEL

WAS

BEGUBNUGN

IN

1880

WILLIAM SHARP

Rightly proud to be a pioneer of the great venture, Mr. Sharp had sought immortality. Unfortunately he had stumbled over the word 'begun' and, unable to erase his false start, compounded his error and left it for posterity.

After some 800 feet had been bored – mainly by Welsh miners – this gallery was left half finished and another shaft, No. 2, was sunk at Shakespeare Cliff on 14th February, 1881. This heading was started under the foreshore, at the same time as the French Tunnel Company was sinking its large main shaft, 18 feet in diameter, at Sangatte, near the point which Low had suggested as the French terminal. Both pilot tunnels, each seven feet in diameter, were pushing forward to meet in mid-Channel.

An interesting feature of Watkin's plans was an alternative to a four-mile-long access tunnel from the South-Eastern Railway line to the Tunnel portal. This took the form of a giant hydraulic lift to carry an entire train down to the Tunnel for its onward journey to France, reversing the process for trains from France. A vast square lift-shaft, 160 feet deep, was to be dug at Shakespeare Cliff, with an underground railway station with electrically-lit waiting and refreshment rooms at the bottom. Absurd as this idea sounds today, it would have been technically feasible, for the Victorian engineers were experts in the application of hydraulic power, and large hydraulically-operated ship docks had already been built in British ports.

Mechanical miracles were the great Victorian vogue, and the compressed air-operated boring machine, invented and constructed by Colonel Frederick Edward Blackett Beaumont, MP, RE, was certainly one of them. One of its major advantages was that it made the use of explosives obsolete. Sir Edward Watkin had sought the

Inscription carved into Abbots Cliff Tunnel by William Sharp in 1880.

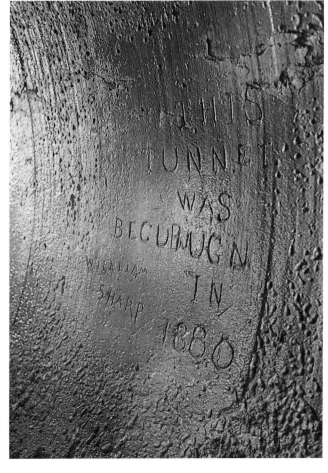

support of Colonel Beaumont, one of the subscribers of the Channel Tunnel Company in 1872, because of his specialised knowledge of tunnelling machinery. In 1874 Beaumont had set his diamond rock drill to work in the Clifton Down Tunnel, with spectacular success – it proved three and a half times faster than manual labour. It was at this stage that another of those 'who invented what first' misconceptions came about. To put the record straight, there is no doubt that Beaumont invented the first rotary boring machine which he patented on 2nd December, 1875. It was, however, Captain Thomas English, of Hawley, near Dartford in Kent, who patented, on 25th October, 1880, the invention of a modified rotary boring machine, which in performance was far superior in every respect to Beaumont's machine. The boring machine which was constructed for the first practical attempt to build the Channel Tunnel was without doubt Captain English's 'improved version', capable of cutting nearly half a mile a month.

Schematic showing shaft to Beaumont's Tunnel from the working site at the bottom of Shakespeare Cliff, Dover.

Plans of Captain English's rotary boring machine, 1880.

The Editor of *The Engineer* credited the machine to Beaumont and, despite a letter of protest from English in May 1883, the Editor did not admit, or correct, the mistake. Beaumont also did nothing to prevent this false impression continuing, as it does to the present day, with frequent references to the 'Beaumont machine'.

POLITICAL AND MILITARY UNCERTAINTY

In June 1881, Sir Edward Watkin told the South-Eastern shareholders' meeting that he expected the seven-foot pilot tunnel to be completed within five years, which was also the calculation of the French Tunnel Company. This gave the Board of Trade food for thought; the date seemed a little too close for comfort, and so the Board resorted to the traditional method of shifting the responsibility for an awkward decision – it suggested to the War Office the setting up of another commission of enquiry.

The new Commission began its meetings in August 1881, with Colonel J. H. Smith, RE, representing the Army and Vice-Admiral Phillimore the Navy. A number of experts and prominent military personalities were also asked for their views.

Further important developments took place in 1881. Sir Edward Watkin, a Liberal MP since 1874, had succeeded in sponsoring a new Parliamentary Act for his South-Eastern Railway, empowering it to go ahead, up to a fixed limit, with tunnelling operations; but now he needed much more money.

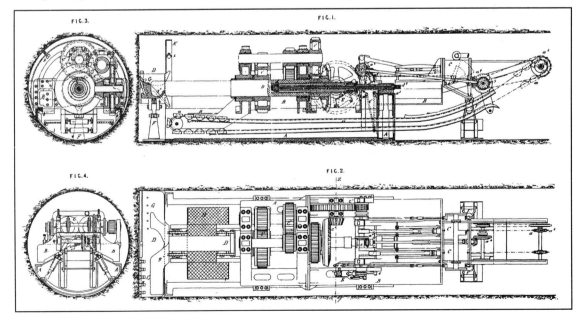

Boring machine in operation, circa 1880.

In September 1881 Watkin applied to the President of the Board of Trade, Joseph Chamberlain, for public funds to bore the 11-mile stretch up to the meeting point with the French in mid-Channel. Still nervous about the speed of developments, Joseph Chamberlain refused his application, and so Sir Edward decided to form a new limited company. It was registered in December 1881, under the name of the Submarine Continental Railway Company, with a capital of £250,000 in £1.00 shares. It took over the rights and existing shafts and headings from the South-Eastern Railway by an agreement signed in January 1882, and work proceeded with renewed vigour – still according to William Low's plans. Watkin offered Hawkshaw and some other members of the Channel Tunnel Company, now more or less inactive, a form of collaboration, but this was declined; instead, the Company prepared its own Bill to be put before Parliament. This was a fatal move.

Sir Edward was also preparing another Bill to enable his new Company to continue beyond the limit fixed by the 1881 Act, having been convinced by the speed of operations that the pilot tunnel could be completed within five years. In April that year he had invited the members of the *Société du Chemin de Fer Sous-marin* for a private visit to the works at Abbots Cliff and Shakespeare Cliff, and they had expressed great satisfaction.

The chief witness to the new military Commission was Lieutenant-General (later Field-Marshal) Sir Garnet Wolseley[15], who submitted

---

[15] Born in Dublin in 1833, Wolseley had a long career of active service behind him, from the Burmese War of 1853 through the Crimea, the Indian Mutiny and Canada to the Ashanti Wars. Gilbert immortalised him as the prototype of his 'Modern Major-General' in *The Pirates of Penzance* in 1880.

his first memorandum in December 1881.

> No question of such vital importance [he wrote] has ever before confronted the British nation . . . the hour when the scheme was definitely accepted would be calamitous for England. The abundance and intricacy of military precautions recommended for safeguarding the Tunnel mouth at Dover were, in themselves, the most convincing proof of the danger hanging over the island nation. No matter what fortifications and defences were built, there would always be the peril of some Continental enemy seizing the Tunnel exit by surprise, and all the commercial advantages of the Tunnel could not outweigh the risk.

Sir Garnet Wolseley, the project's fiercest critic.

General Sir John Adye, Surveyor-General of the Ordnance, in his own report, submitted to the Commission in January 1882, disagreed with Sir Garnet:

> A French strategist who wanted to invade England would never regard the Tunnel as the answer to his problems. He would in any case have to mass his forces first on the French coast, and if Britain's intelligence service was of any consequence it would

in good time give a warning of such troop concentrations so that adequate measures could be taken. But even if invading units started moving through the tunnel, a small force of infantry and artillery could destroy them with ease as they emerged at Dover. As to the danger of an invading force seizing the Tunnel exit, if it really succeeded in establishing a bridgehead in England it would not be likely to turn aside to capture such an uncertain line of communication as the Tunnel [probably flooded by that time], but would advance on London, provided it had been sufficiently reinforced for such attack. And what [Sir John asked] would the Royal Navy be doing during the time the invaders were massing in France, crossing to England, and establishing their bridgehead?

Somewhat surprisingly, it was a naval officer who was sceptical of the Navy's defensive role A letter written by Admiral Cooper Key to the First Lord of the Admiralty was submitted to the Commission:

> Any machinery for destroying the Tunnel in an emergency could not be relied upon. After seizing the Tunnel, the enemy would march on London, with the Navy a powerless spectator.

The situation was exacerbated by political instability in France, which culminated in February 1882 with the resignation of the French Prime Minister Gambetta. Seven years earlier the French Republic had received its parliamentary sanction, but France was far from settling down under a constitutional regime. The old war hero Marshal de MacMahon, replacing the moderate Thiers as second president of the Third Republic, was expected to keep the parties reasonably at peace; but he was himself suspected of harbouring plans for a restoration of the monarchy. Cabinets changed frequently; there was more than one attempt at a *coup d'état*, and the aged Marshal resigned early in 1879. Against this background of internal strife, France's foreign policy and international involvements appeared uncertain and in considerable disarray. Disraeli took advantage of the French vacillations in 1875 to acquire for Britain 176,000 of the 400,000 Suez Canal shares, which created a great deal of anti-British feeling. As Anglo/French relations deteriorated, the French Government, in 1878, cancelled the Anglo/French commercial treaty of 1860.

Gambetta had tried in vain to mend matters; and spoke in the Chamber prophetically about the European role of the two countries in western politics: 'I cannot imagine any other European policy [than that of Anglo/French friendship] which could help us in the difficult circumstances that may arise.' He was, however, unable to form a cabinet, and so his vision of a close collaboration between France and Britain faded – as had British hopes of avoiding a trade war with France when the commercial treaty was

Artist's impression of Sir Edward Watkin defeating French invaders by opening the Tunnel's sluice gates.

cancelled. Tariff barriers went up, and the age-old English suspicions of her neighbour across the Channel reached new heights.

Thus, within five years of the Anglo/French agreement on a Tunnel Treaty, the response of the British authorities and general public veered from sympathetic approval, through lukewarm interest, to downright opposition motivated by an almost superstitious fear.

That same month the Commission informed Chamberlain and the War Office that it could not come to any final conclusions on defence matters regarding the Tunnel. So yet another commission was set up, a 'Military Committee', under the chairmanship of Major-General Sir Archibald Alison and including, among its eight members, other top-ranking officers, the Home Office Inspector of Explosives, the Chief Engineer of the General Post office, and Professor Abel, Chemist to the War Department.

Sir Edward Watkin must have been rather alarmed by the composition of this new Committee. These were not men renowned for liking the Tunnel scheme. At the same time, an anti-Tunnel petition was circulating among influential people, and the journal *Nineteenth Century* acted as the rallying point of the campaign. Alfred Lord Tennyson, T. H. Huxley, Robert Browning, Herbert Spencer, Lord Dunsany and Sir George Sitwell signed the petition against the Tunnel, though their reasons and attitudes were by no means uniform.

In its issue of 4th March, 1882, the *Illustrated London News* published four pages of pictures and a report of the first two press and VIP visits to the Channel Tunnel. The pilot tunnel from Shakespeare Cliff was already well out under the sea for a distance of nearly three-quarters of a mile, as was the French heading.

> On Saturday the 18 (February), Sir Edward Watkin conducted a party of thirty or forty gentlemen from London to inspect these works, the Lord Mayor of London being one of the party. They descended the shaft, walked a thousand yards under the sea, and admired the working of Col. Beaumont's compressed-air boring machine. They had the electric light, by which the tunnel was illuminated from end to end. In anticipation of this visit Sir Edward had directed a luncheon to be prepared in the tunnel, which was partaken of in a chamber cut in the side of the heading, tables and stools being set there for the occasion . . .
>
> The Channel Tunnel was again opened to another party of London visitors on Tuesday of last week . . . Under the guidance of Mr. Francis Brady, C.E., engineer of the Channel Tunnel, and Colonel Beaumont, R.E., the visitors, six at a time, having put on rough overalls to save their clothes from dust,

descended into the shaft by means of an iron cage, such as is
used in coal-mines. The shaft is sunk in the chalk cliff at the
foot of the 'Shakespeare Cliff', between Folkestone and Dover,
and is about one hundred and sixty feet in depth. The opening
is circular, with boarded sides, and the descending apparatus is
worked by a steam-engine. At the bottom of this shaft is a
square chamber dug in the grey chalk, the sides of which are
protected by heavy beams; and in front is the experimental
boring, a low-roofed circular tunnel, about seven feet in
diameter, the floor of which is laid with a double line of tram-
rails. This tunnel is admirably ventilated, and on visiting days is
lighted with electric lamps, the steam-power at the mouth of the
shaft being sufficient for all purposes. The stratum through
which the experimental borings have been made is the lower
grey chalk. This material, while perfectly dry, and very easily
worked, is sufficiently hard to dispel any apprehensions of
crumbling or falling in . . .

The length of the Submarine Continental Railway
Company's Tunnel, under the sea, from the English to the
French shore, will be twenty-two miles; and, taking the shore
approaches at four miles on each side, there will be a total
length of thirty miles of tunnelling . . . The shaft goes down to
the beginning of the tunnel, which is here 100 feet below the
surface of the sea . . . A heading, now three quarters of a mile
long, has been driven in the direction of the head of the
Admiralty Pier [Dover], entirely in the grey chalk, near its base,
and a few feet above the impermeable strata formed by the
gault clay . . .

The present heading is 7 feet in diameter. Machinery is
being constructed by which this 7 feet hole can be enlarged to
14 feet by cutting an annular space, 3 feet 6 inches wide,
around it. This will be done by machinery . . . furnished with
an upper bore-head . . . One machine will follow the other, at a
proper interval; and the debris from the cutting by the first will

be passed out through the second machine. The compressed air, likewise, which is necessary to work the advanced machine, will be similarly passed through the machine coming behind . . . only two men are at present needed for each machine.

At the end of the tunnel the visitors found one of the Beaumont compressed-air boring machines at work. The length of this machine from the borer to the tail end is about 33 feet. Its work is done by the cutting action of short steel cutters fixed in two revolving arms, seven cutters in each, the upper portion of the frame in which the borer is fixed moving forward 5-16ths of an inch with every complete revolution of the cutters. In this way a thin paring from the whole face of the chalk is cut away with every turn of the borer . . . A man in front shovels the crumbled debris into small buckets, which, travelling on an endless band, shoot the dirt into a 'skip' tended by another man. The skip when filled is run along a tramway to the mouth of the shaft. At present these trolleys, each holding about one third of a cubic yard, are drawn by men; but before long it is hoped that small compressed-air engines will be used for traction. The rate of progress is about one hundred yards per week, but will soon be much accelerated. As worked at present, the number of revolutions it makes is two or three per minute, which . . . amounts to bring nearly an inch a minute while the machine is at work. But Colonel Beaumont anticipates no

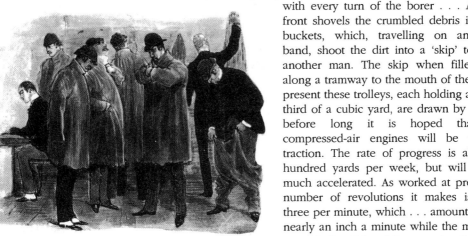

difficulty in making the machine cut its way at the rate of 3-8ths of an inch per revolution, and getting five revolutions per minute, which would give a rate of advance of two inches per minute.

A very important question has been raised with regard to the supply of compressed air. Carried in four-inch iron pipes, it now reaches the machine with a pressure of about 20 lb, the pressure at the compressor at the shaft mouth being from 30 lb to 35 lb; but by increasing the diameter of the supply pipe to eight inches the loss of working value by friction would be greatly diminished, if not rendered inappreciable. The boring has now advanced to the length of 1,250 yards . . . and it is going on at the rate of three miles a year, which speed of working, as we said, will be increased. Simultaneous borings from the French side at the same rate would give six miles a year, or a complete tunnel underneath and across the Channel in three years and a half . . .

The shape which the completed tunnel will assume will probably be a circle, 14 feet in diameter, but flattened at the bottom to receive the rails. It will be lined with two feet thickness of cement concrete; not that this is necessary to ensure the stability of the work, but to prevent accidental falls of chalk. The concrete will be made of shingle from Dungeness, and of cement formed from the grey chalk excavated from the tunnel itself. In this manner, the tunnel will afford the means of its own lining at a cheap rate. The gradients will be 1 in 80, on each side, until the depth of 150 feet below the bottom of the sea is reached; after which the line may be said to be level, subject only to a very slight inclination from the centre outwards, to prevent the lodgings of water.

The ventilation of the tunnel is, perhaps, the simplest matter in connection with it, but as some doubts have been expressed upon this, it may be here shortly explained. During the construction of the tunnel, the air necessary for ventilation will be more than enough supplied by that used to drive the boring machines . . . When the tunnel is opened for traffic, the trains will run through by means of Beaumont compressed-air locomotives . . .

The Channel Tunnel locomotive will weigh from sixty to seventy tons, and will be charged with 1,200 cubic feet of air, compressed to the density of seventy atmospheres, the equivalent of which is over 80,000 cubic feet of free air. This will give power sufficient to draw a train of 250 tons gross weight (including the engine) the distance of twenty-two miles under the sea. Assuming that the rate of travelling be thirty miles an hour, the air discharged by the engine would give a supply of free and pure air to the amount of 2,000 cubic feet, approximately, which will be far in excess of what is needed by the passengers in the train . . . Reservoirs will be placed at convenient intervals, so that the engines, should they need it, may be replenished with compressed air. It will, therefore, be seen that Colonel Beaumont's system of compressed-air engines

affords equal advantages with the ordinary steam locomotives, and with no increase in weight.

One of the illustrations shows 'how the tunnel is defended by existing works' and the last part of the article was entirely concerned with defence:

> The controversy now going on between different military authorities and politicians respecting the effect which the Tunnel would have upon our insular safety from the risk of a foreign invasion has already been much noticed. In the *Nineteenth Century* for March, Colonel Beaumont replies to the arguments of Admiral Lord Dunsany and of a distinguished military man, understood to be Sir Garnet Wolseley, who disapprove of the Tunnel upon this ground. Having been himself three years in the construction of the Dover fortifications, Colonel Beaumont is enabled to assure us that by the natural strength of the position, and by the powerful works erected there . . . Dover may be regarded as a 'first-class fortress, quite safe from any *coup de main* from without' . . . There will be arrangements, under control of the military for letting the water of the sea into the Tunnel; but these arrangements, which will be kept secret, will be of such a nature that they cannot be tampered with improperly, while they can be promptly put in operation without the assistance of technical experts. The position of the inclined gallery, connecting the end of the Tunnel with the main railway lines, will be such that the trains, on emerging from under the sea, must be lifted bodily, by suitable hydraulic apparatus, to the daylight surface; and without the aid of such hydraulic apparatus, the ends of the tunnel will be blocked . . . Hence it will be evident that, supposing a party of two thousand men could pass through the Tunnel by surprise, and could reach the bottom of the shaft at the Dover end, they could surely get no further . . .
>
> But Colonel Beaumont does not admit that it is possible for a surprise party of two thousand men, as imagined by Lord Dunsany and his military authority, to pass through the tunnel unobserved. They cannot come by trains; as, irrespective of any suspicions on the part of the booking clerks, special train arrangements would have to be made to carry so large a number; they cannot march, as they would be run over by the trains, running as they will, at intervals of ten minutes, or oftener, without cessation, day or night . . . He [Beaumont] thinks arrangements should also be made by which the ventilating

engines, used for the ordinary purposes of the tunnel, could pump the smoke from their own furnaces into the Tunnel, in place of fresh air; this . . . would soon produce an atmosphere through which no living being could pass . . . If we had lost command of the sea temporarily and the enemy had landed twenty or thirty thousand men on our coast, there would still have been time for us to block or flood the Tunnel, or to destroy its ventilation; at any rate, to destroy the hydraulic lifts, which could be done by firing a single charge of dynamite . . . It would appear, therefore, that the only time when, by any stretch of imagination, the tunnel would be a source of danger, no invader could by any possibility make use of it.

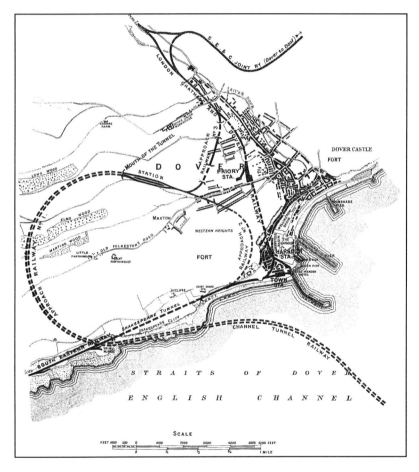

Plan of Dover circa 1880 showing Shakespeare Cliff and the route of Beaumont's Tunnel.

Watkin laid on a few more VIP visits to the workings[16], with special trains, lavish buffets, and dinners at the Lord Warden Hotel in Dover, where rousing speeches in support of the Tunnel were made. A fortnight after the press visit described in the *Illustrated London News*, a further seventy guests were taken down into the

---

[16] Watkin had installed strings of incandescent electric lamps, invented three years earlier by Joseph Wilson Swan.

pilot tunnel, among them the Prime Minister and Mrs. Gladstone, the Prince and Princess of Wales, the Dukes of Cambridge and Sutherland, the Archbishop of Canterbury – and Sir Garnet Wolseley himself who, nevertheless, did not change his mind about the perils of the Tunnel.

Visit of military officers to the Channel Tunnel works, 1882.

In May 1882, the Military Committee
delivered its report to Parliament, stipulating
its conditions for the defence of the Tunnel –
or rather its destruction. All surface works
should be commanded by the advanced
works of a new fortress and permanently
mined; there should be a portcullis for
closing the Tunnel mouth and an apparatus
for releasing asphyxiating gas against an
invader; sluice gates and explosive charges
for flooding and blowing up the galleries
should be installed, and all mines should be
controlled from the fortress as well as from
points further inland. Watkin had no
objection to this: 'You may touch a button at
Horse Guards and blow the thing to pieces',
he had told the Committee. But neither of
the two Bills which the Channel Tunnel

Company and the South-Eastern had placed before Parliament
received the Committee's recommendation. The portents for the
Tunnel were grave.

The *coup de grace*, when it came, did not come from the
military, at least not directly. It came from the Board of Trade.

In mid-January the Board had drawn Watkin's attention to
section 77 of the South-Eastern Railway Act of 1881. This section
concerned the foreshore rights, forbidding any tunnelling beyond
the low-water mark without explicit permission. Early in March, the
Board wrote again, reminding him that he would be violating the
Crown's foreshore rights if he were to continue digging into the
three-mile limit beyond high water. Watkin replied that he was
driving his pilot tunnel out to sea without infringing the rights of
the Crown as the land from which he was operating had been
purchased from the Church of England and therefore, by ancient
traditional rights, no longer belonged to the Crown.

The Board of Trade did not dispute this; but replied that the
Crown retained control over the waters between high tide and the
three-mile limit. Assuming that Watkin's heading must already have
reached the limit permitted by the 1881 Act the Board demanded
that he should now stop. Again Watkin took no notice. The Board
sent investigators to the Shakespeare Cliff workings to inspect what
was going on; and the Secretary of the Submarine Continental
Railway Company agreed to stop operations as the Board of Trade
demanded. Watkin sent a telegram to the President of the Board,
Joseph Chamberlain, on 9th April, informing him that the Beaumont
machine had stopped, but as the ventilating air was transmitted by
the machine there was concern about the atmosphere in the

heading. Chamberlain replied, also by telegram, that the machine would be allowed to go on working if workmen's lives were in danger, and that he would order an investigation whether there was any need for workmen to remain in the heading.

As a result, Colonel Yolland, RE, Chief Inspector of Railways to the Board of Trade, was told to investigate. Watkin replied that no visits could take place for another week as the Duke of Edinburgh was expected to come on 18th April. Chamberlain was angry: what did Colonel Yolland's inspection have to do with the Duke's visit? Watkin replied that he would be happy at any time to welcome Mr. Chamberlain personally. The latter declined the offer, but the jostling for position continued. Watkin, backed by his fellow-directors, kept the Board of Trade at bay. There was always some reason, during the next three months, why Colonel Yolland could not be received or make his inspection. Meanwhile, the boring machine continued to operate, excavating the chalk, watched by various groups of visitors – none, however, from the Board of Trade.

Gradually, the correspondence between the Board and Watkin grew more acrimonious. Terms such as 'insinuations without foundation', 'deliberate personal insult', 'needless and undeserved threat' abounded, and eventually both sides declined to communicate. Early in July the Board applied for an order from the High Court of Justice, which it received on 5th July, giving legal access to the No. 2 (Shakespeare Cliff) heading. The implication was obvious: the digging had to stop.

Colonel Yolland went to the Shakespeare Cliff workings on 8th July, armed with a copy of the injunction but with little else in the way of equipment. A week later, on 15th July, he returned with measuring equipment, and found that more than 1,800 feet of Crown property had been undermined by the pilot tunnel, that the heading could have been ventilated directly without channelling the compressed air through the Captain English boring machine, and that provisions for removing ground water and for sheathing fissures in the gallery walls were, in his opinion, inadequate. His critical mood was not exactly soothed by an unfortunate incident – he slipped on wet ground in the heading, fell on his back, and bumped his head against a truck rail. In his report to the Board, 'Colonel Yard' (as the workmen in the Tunnel had nicknamed that measuring gentleman) recommended that Watkin's Submarine Continental should be forced to stop operations at once.

However, when no visitors were about operations continued at full steam. This is evident from a work sheet, probably by the hand of the supervising engineer, covering the last period of work in 1882; it came to light ninety years later in British Rail's archives. It is headed 'Channel Tunnel Works – Experimental Boring (seven feet diameter) in an easterly direction from No. 2 Shaft, near the western end of Shakespeare Tunnel, Dover', and records 'lengths of heading bored from week to week from April to July 1882. During the fourteen weeks recorded, 1,888 feet of heading were bored, an average of just over 106 feet per week.

The most interesting feature was in the last column of the sheet, headed 'Remarks'. The greatest length bored in any week was 173 feet, ending on 20th April, although work was 'delayed by visitors, by fixing tubbing and by moist chalk adhering to buckets'. The week after: 'Hydraulic jacks gave way'. Then again, 'Delayed by repairs, renewals and visitors'. One day there were 'not sufficient hands'; and on another day 'hydraulic jacks gave way again'.

Work progress sheet, 1882.

Channel Tunnel Works.

Experimental Boring (7 ft. dia.) in an easterly direction from Nº 2 Shaft, near the western end of Shakespeare Tunnel, Dover.

Lengths of heading bored from week to week from April to July, 1882, by the Beaumont Boring Machine. (Progress not recorded previous to Apl. 13)

Week ending. 1882.	Length of Heading bored during week.	Remarks.
April 20th	173 feet	Delayed by mechanical defects and by visitors.
„ 27	143 „	Delayed by visitors, by fixing tubbing and by moist chalk adhering to buckets.
May 4	131 „	Hydraulic jacks gave way.
„ 18 (Fortnight)	144 „	Delayed by repairs, renewals and visitors.
„ 25	95 „	No delays.
June 1	76 „	Not sufficient hands one day; hydraulic jacks gave way on another.
„ 8	104 „	No delays.
„ 15	110 „	Delayed by visitors.
„ 22	109 „	No delays.
„ 29	112 „	No delays.
July 6	106 „	Delayed by visitors.
July 13	87 „	Hydraulic jacks gave way on 7th & 11th; delayed by visitors on 8th July.
July 20	98 „	Delayed by visitors and by Board of Trade inspection on 15th July.
14 weeks	1,488 feet.	106·3 feet = average progress per week.

For three weeks there were 'no delays', but then work was again 'delayed by visitors'. In the week ending 13th July, only 87 feet were bored: 'Hydraulic jacks gave way on 7 and 11; delayed by visitors on 8th July'. There followed the last week, ending 20th July: 'Delayed by visitors and by Board of Trade inspection on 15th July'. But all the same 98 feet was bored before the digging stopped.

By the end of that fateful Summer of 1882, the Submarine Continental Railway Company had tunnelled 897 yards from Abbots Cliff – seven feet in diameter – parallel with the shore below and the Dover-to-Folkestone railway above. The No. 2 main pilot tunnel, the Shakespeare Cliff heading, was 2,040 yards long with its working face 130 feet below the sea bed, running almost parallel with the coastline for a few hundred yards, then reaching out to sea towards the Admiralty Pier, Dover. The No. 3 heading at the eastern end of the Shakespeare Tunnel, only four feet in diameter was 85 yards long, and it was intended as an air and pumping gallery, with a cross-gallery linking it to No. 2.

Altogether, the Submarine Continental Railway Company had spent about £100,000. The Captain English machine had reached

Workshop for the compressed-air machines at Sangatte, 1880s.

the highest expectations, and the influx of water was minimal, requiring only half a day's pumping once every two weeks.

The French, understandably, were puzzled by the Board of Trade's injunction. Surely that absurd security scare was only a temporary diversion and in a few weeks the digging would restart? But even Sir Edward Watkin, usually the greatest of optimists, could offer no comfort. On 18th March, 1883, the French *Société Concessionaire du Tunnel sous la Manche* – as it was now called – also stopped work, having completed 1,840 metres, a length similar to the English heading. It had been using the Captain English machine, constructed by the *Société de Construction de Batignolles* which made 100 revolutions per minute, but progress had been hampered by more difficult conditions. The French shaft was to remain closed for a very long time. The English workings, however, were kept open. Watkin insisted on continuing with the maintenance work, and still retained some staff. Colonel Yolland paid the Tunnel another visit early in August, did some measuring, and reported to the Board of Trade that in gross defiance of the injunction a further 70 yards had been dug since his previous visit in July. The Board of Trade acted on his report and brought a new court action against the Submarine Continental Railway Company, but chief engineer Francis Brady claimed that only ventilation work had been carried out, to ensure that the Colonel did not suffocate during his forthcoming visit! Yolland paid another visit at the end of August which resulted in his admitting to the Board that he had miscalculated.

Though it was not yet recognised, the nineteenth century project to build a Channel Tunnel had received a mortal blow.

CHAPTER THREE

BLOODY BUT UNBOWED

1882 – 1974

THE LANSDOWNE COMMITTEE

Between late 1882 and early 1883 the rival Tunnel Companies brought Bills before Parliament and, once again, it was up to Westminster to take a decision. The President of the Board of Trade, Joseph Chamberlain, decided that a Joint Parliamentary Select Committee, comprising five members from the Lords and five from the Commons, should be set up to hear evidence on the Tunnel and pronounce a verdict.

Before the proceedings started, the London, Chatham and Dover Railway Company fired a broadside at its rival by submitting a petition to the Commons. It complained that the South-Eastern Railway Company's Bill threatened to infringe its rights and interests and 'humbly prayed that the Bill may not pass into a law as it now stands'. To be sure, Sir Edward Watkin had done his best to annoy and provoke James Forbes, the Chairman of the London, Chatham and Dover Railway Company, and Lord Grosvenor, still head of the Channel Tunnel Company. There had been an angry exchange of correspondence between them in 1881-82. Watkin had acquired a private letter from Grosvenor to Forbes and exploited it unscrupulously; the letter was more or less a plan for concerted action against Watkin's company, although Grosvenor explained that 'never having been able to obtain your co-operation in any way, I naturally have turned to Mr. Forbes'. Watkin also ridiculed Grosvenor's Tunnel plan by telling him that it merely showed a route beginning at Fan Hole on the coast and ending at Biggin Street in Dover, without any link either with the Chatham or the

South-Eastern line: 'You have deposited no plan nor made any money deposit for anything more', Watkin wrote. No wonder Grosvenor disliked him, but his petition, submitted in February 1883, did neither his own company nor the Tunnel scheme itself any good; it merely confused the issue.

The Joint Select Committee, under the chairmanship of Lord Lansdowne, first met on 20th April, 1883, and after 14 meetings, ended on 21th June. The Committee's brief was 'to enquire whether it is expedient that Parliamentary sanction should be given to a submarine communication between England and France'. Provided the answer was yes – even subject to conditions to be imposed by Parliament – the real question to be answered was whether the economic advantages of a Tunnel would outweigh its strategic disadvantages.

The Committee summoned forty witnesses and asked 5,396 questions. Among the witnesses were Sir Edward Watkin, Lord Grosvenor, Sir John Hawkshaw, Colonel Yolland, Colonel Beaumont, Francis Brady, James Forbes, the Duke of Cambridge (Commander-in-Chief of the Army since 1856 and cousin of the Queen), and a number of prominent men in trade and industry, as well as a few generals and admirals. The last of the military experts to be called as a witness was Lieutenant-General Sir Garnet Wolseley.

When the report of the Select Committee was published it was not the overwhelming advantages to Britain's trade and industry which claimed public attention, but Wolseley's evidence, which raised fears of invasion by a foreign power via the Tunnel to an unprecedented height.

Wolseley stressed all the familiar military objections, already raised by the Duke of Cambridge, but the General was riding his hobby horse and gave full rein to his passionate loathing of the project.

The General revealed himself to be a formidable demagogue with his facile jibes at people's 'mercantile activity', the 'hunt after riches', at 'railway speculators' and 'selfish cosmopolitans', and his spectres of higher taxes and periodical panics: what would it all be in aid of? To save a few tourists from 'a little sea-sickness'; what of the horrors of invasion? His harping on the general fear of sea-sickness, which to his mind was ridiculous, gave him away, as did his absurd claim that steam-power at sea had been 'the first great step towards the destruction of our former naval supremacy'. The General was not only a demagogue, he was also something of a crank.

Although he won support from the uninformed die-hards and chauvinists, he did not convince the Select Committee, whose members sought to reach a reasonable answer. Several conflicting

Alarmist reaction: Sir Garnet Wolesley flees the French invader (the American Puck, *1887).*

draft reports emerged, none of which received the 'entire approval' of a majority. Lord Lansdowne's own report concluded: 'We have no course open to us except to recommend that this enterprise should not be prohibited on merely political grounds, and that it be allowed to proceed.'

Lansdowne held a minority view. By a majority of six to four the Committee reported that '. . . it is *not* expedient that Parliamentary sanction should be given to a Submarine Communication between England and France'. The Report was adopted on 10th July, 1883.

A few weeks later the Bills before Parliament were thrown out. The South-Eastern and Submarine Continental Railways introduced a Bill the following year. This, too, was opposed by the Government and defeated on 14th May, 1884. However, 84 Members supported it. The Channel Tunnel Company introduced two more Bills in 1885 and 1886; the first was defeated and the second was withdrawn as a General Election was pending.

THE RIVALS MERGE

On 8th July, 1886, in the Board Room at London Bridge Station, the Submarine Continental Railway Company held an Extraordinary General Meeting which passed the resolution:

> That the capital of the Company be increased by the sum of £25,000, in 25,000 fully paid-up shares of £1 each, for the special purpose of the purchase of all the rights and properties of the Channel Tunnel Company and its shareholders.

Negotiations had been going on for some time. Each Company had realised that its rivalry, particularly in the face of the opposition it had provoked, was ridiculous, if not self-destructive. A merger was the answer, and as the Submarine Continental Railway Company could raise the money, it was decided that it should buy the Channel Tunnel Company at a bargain price: its only assets, apart from its name, were two derelict shafts and a bit of land at St. Margaret's Bay. On 24th February, 1887, another Extraordinary General Meeting passed the resolution 'that the name of the Submarine Continental Railway Company Limited be in future The Channel Tunnel Company Limited'. A further meeting, on 11th March, confirmed this change and so *the* Channel Tunnel Company[17] which exists today, and whose shares are still quoted on the London Stock Exchange, came into being.

The new Company's first step was to introduce, in August 1887, a new Bill into Parliament to enable it to continue its

---

[17] Renamed 'Channel Tunnel Investments' on 15th June, 1971.

experimental works. Simultaneously it launched a public petition in support of its 'commercial enterprise of vast mercantile value, and an international work which cannot fail to promote peace'. It was signed by nearly fifty prominent personalities in politics and commerce. New names were appearing in connection with the project, including City banker Baron Frederick Emile d'Erlanger, formerly a director of the old Channel Tunnel Company and soon to become Chairman of the new company's Board. Francis Brady was now the leading engineer, while the elderly Sir John Hawkshaw was fading into the background.

GLADSTONE AND THE TUNNEL

> Happy England that the wise dispensation of Providence has cut off by that silver streak of sea which passengers so often and so justly execrate – partly from the dangers, absolutely from the temptations which attend upon the local neighbourhood of the Continental nations.

The old village of Sangatte, prior to the arrival of Tunnel workings.

So spoke Gladstone in 1870, the year of the Franco/Prussian War. The Channel, he argued, saved England from the 'miserable burden of conscription'. However, like most politicians, Gladstone was not incapable of changing his mind.

Sir Edward Watkin was a shrewd business tycoon and Gladstone soon succumbed to his charms, becoming a close personal friend and a Tunnel enthusiast. He strongly opposed the military which had brought the notion of the Tunnel's threat to National security to fever pitch. Watkin invited him to inaugurate the construction of Hawarden Bridge on the Chester-Wrexham line in 1887 and in his address Gladstone said:

> Sir Edward Watkin is one of those men wicked enough to desire that a Tunnel should be built under the Channel to France; and what is truly painful to me is that I am compelled to confess before you – and I do it publicly – that I am one of those men who are wicked enough to agree with him!

As the introduction of Tunnel Bills into Parliament, and their inevitable defeat, became routine, Gladstone, as Leader of the Opposition, would speak in its favour. His longest and wittiest Tunnel speech was that of 5th June, 1890, at the second reading of the Channel Tunnel (Experimental Works) Bill. Having re-affirmed his continuing support for the Bill and for the project he praised the French for their forbearance:

> I may also add that, whilst I think that our position in respect of the Government of France on this question of the Channel

Tunnel is a humiliating position, on the other hand, the Government of France deserves, in my opinion – and I am glad to take this opportunity of declaring it in this House – the highest credit and the warmest acknowledgement on our part for never having made our altered position the subject of complaint [*Cheers*] . . .

In his rebuttal of Hicks-Beach's claim that the commercial advantages were but 'vague expectations' he said:

Has the Right Hon. Gentleman ever read any examinations of the witness for the first projects of railways in this country? Does he know that George Stephenson was challenged boldly and most confidently to say whether he would undertake to give his judgement that the steam engine would be able to drag a train of carriages at ten miles and hour? And, further, he was pressed as to the possibility of eight miles an hour; and finally, whether he would guarantee that the train would go at four miles an hour. [*Hear, hear*] . . . I have great respect of professional authorities, but with regard to the amount of danger – and that distant danger [of invasion by the tunnel] – to be incurred, I do not think that they are in any degree to be considered as the best authorities. [*Hear, hear*] At this moment my belief is that the people of England are not opposed to this Tunnel. [*Cheers*]

He dealt simply and succinctly with the fear of invasion by putting it into historical perspective:

I believe that we have invaded France ten times for once that France has invaded us. We have held the capital of France alone once, and we have entered it in conjunction with other Powers, and if there is a country which would be justified in feeling sore and apprehensive on the subject of the Channel Tunnel, that is the French nation. In France there has been no apprehension. The French know that we are masters of the sea, and if we were to cease to have possession of the Channel that would, for the purpose of invasion, be fatal to our position. It does not turn upon the Channel Tunnel in the slightest degree . . .

Gladstone concluded as he had begun:

I must repeat the sentiment which on every occasion I have been ready to express, and say that I believe this [the Channel Tunnel] to be a considerable measure and a useful measure, and that the arguments opposed to it deserve neither acceptance nor respect. [*Cheers*]

But the Bill was again defeated, this time by 81 votes – as was the next Tunnel Bill two years later. It seemed that its opponents always won the day. 'So long as the ocean remains our friend', said Arthur Balfour, 'do not deliberately destroy its power to help us.' And Lord Randolph Churchill, in a bitter attack on the 'mischievous project', called upon the House to keep England, 'as it were, *virgo intacta*'. The House complied.

NEW SCHEMES

In the 1830s Thomé de Gamond had examined the idea of spanning the Channel by a bridge and rejected it. Subsequent bridge proposals did not get much further because of their cost and danger to shipping; but engineers and industrialists kept returning to the bridge concept, which had the effect of frustrating the Tunnel's advance.

The best-designed bridge and the most weighty – in every sense – was that of the Channel Bridge Company, formed in 1884 under the auspices of Schneider, the French armaments king who had supplied the country's armed forces with military hardware for half a century. As it seemed unlikely in the 1880s that a major new steel-consuming war would break out, Schneider responded favourably when, Hersent, ex-President of the French Civil Engineers' Society, came to him with the idea for a Channel Bridge with the glowing prospect of a steel structure 24 miles long. Sir Benjamin Baker, who was just completing the first railway bridge across the Firth of Forth, and his collaborator Sir John Fowler, both eminent British engineers, were commissioned to work out a blueprint for the bridge together with Hersent.

The railway-only bridge would span the narrowest point of the Straits, between Cap Blanc Nez and South Foreland. Its 72 steel piers resting on massive masonry support, would carry the platform nearly 180 feet above high water – more than enough headroom for even the largest ships of the day. The cost would be about £34 million, or three times the cost of a Tunnel at the time. The plans were displayed at the 1889 Paris Exhibition, and presented to the Iron and Steel Institute in Paris. The French Government, however, advised by its experts, rejected the Channel Bridge Company's request for a concession; not only because of the danger to Channel shipping, or the cost of the scheme, but also because knowledge of the sea bed was limited and the construction of the piers extremely hazardous.

Other related schemes were bound to follow, such as the *passage mixte* – or the bridge-tunnel-bridge project – by the French goldsmith and engineer, Philippe Bunau-Varilla, which envisaged two monster elevators lifting and lowering the trains between the mid-Channel bridgeheads and the tunnel itself.

By this time Sir Edward Watkin had become eccentric, though

Artist's impressions of Schneider/Hersent Channel Bridge Project, 1891.

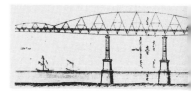

his energy did not seem to wane with increasing age. He still hoped to extend his railway empire, which stretched from the Welsh coast to the south-east coast, to France within his lifetime. Frustrated, he would turn to new ideas. For a while it was towers. Workmen who had been in the No. 2 heading erected a 130-foot tower, made of old rails, above Shakespeare Cliff, topped by a 24-foot-tall lantern which shone across the Straits until it was pulled down in 1913. He also made a start on a 1,150-foot tower at Wembley Park in Middlesex, designed to rival Eiffel's tower in Paris.

Coal provided a curious dimension to the Tunnel story. Some geologists had suggested that there was a deep coalfield in the Dover/Folkestone area extending under the sea bed. Watkin and Brady applied to the Department of Woods and Forests in 1886 for a lease of the undersea minerals, but it seems that the tunnellers had irritated the Department, and the lease was refused. The engineers applied on behalf of the Channel Tunnel Company for a lease from the Ecclesiastical Commissioners, but they also declined. Thus the Company was left holding a few acres of Kent around its shafts, sufficient for driving a Tunnel to France but not for the sinking of experimental pits and the erection of minehead workings for the extraction of coal. Watkin insisted on going ahead secretly and in November 1892 Francis Brady carried out test borings. The largest coal seam, four feet thick, was discovered at 2,222 feet, with the result that Watkin was again accused of tunnelling towards France, using his quest for coal as a cover. In 1895 a further attempt was made to win coal by a syndicate of financiers, this time under a lease from the Ecclesiastical Commissioners. In 1895 work was started on the sinking of three shafts – the deepest to the depth of 1,600 foot – and over £150,000 was ploughed into the development. However, no coal seams of any consequence were discovered and the Kent Collieries were eventually abandoned.

In 1897 the capital of the Channel Tunnel Company was reduced from £275,000 to just over £90,000 by Special Resolution. The following year the Chancery Division of Justice, by an action of Her Majesty's Attorney General against the South-Eastern Railway and the Channel Tunnel Company Ltd., ordered that the two Companies 'be perpetually restrained from further boring into or excavating or in any way interfering with the bed of the sea below ordinary low-water mark off the coast of Kent between Dover and Folkestone without the consent of the Board of Trade . . .'

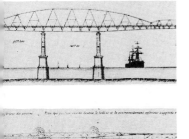

INTERNATIONAL TENSIONS

The political climate between Britain and France became hardly conducive to a renewal of efforts to build the Tunnel. France had been suffering from a series of internal upheavals – the Boulangist movement, with its threat of a *coup d'état* and military dictatorship; the Panama scandal, which landed Ferdinand de Lesseps and his son, though innocent, in jail; and the Dreyfus affair which split the nation. England looked on in shocked surprise and concern: France had become a poor partner in the *entente cordiale* which had seemed to so many Englishmen the only hope for European peace and prosperity. Then came the Fashoda affair of 1898.

A French commander, marching to the Fashoda post in the Sudan with a force from the Congo, had hoisted the French flag, and Sir Herbert Kitchener, who had just captured Khartoum, re-hoisted the British and Egyptian flags, and asked the French to withdraw. They refused, there were angry diplomatic exchanges between London and Paris, and tension rose. In the end the French gave in and withdrew, but much bitterness remained between the two countries. However, after the death of Queen Victoria – whose personal opposition to the Tunnel was no secret – the new King Edward VII, became increasingly disenchanted with the antics of Queen Victoria's grandson, Kaiser Wilhelm II of Germany. Not only did he resolve differences of opinion between France and Britain, but he also established, on a personal and political level, the *entente cordiale* which had been discussed, but never properly forged since Louis-Philippe had coined the phrase in the 1840s. Yet thoughts of a cross-Channel link, as the new century dawned, continued to languish in a slough of despond. All that remained were two short lengths of Tunnel extending below the Channel – one from England and one from France: twin symbols of one hundred years of frustrated endeavour.

Louis Blériot – the first man to cross the Channel by aeroplane.

A NEW DIMENSION

The Industrial Revolution had set the pace with the advent of railways and steam ships but the horse was still the prime mover on the road. However, between the early 1880s and the early 1900s – a span of a mere twenty years – there emerged a spate of exciting and significant inventions which were to revolutionise transport on land, sea, and air. In 1885 the internal-combustion engine was invented by two German engineers, Karl Benz and Gottlieb Daimler – remarkably they achieved this quite independently of one another. France immediately recognised the automobile's potential, unlike Britain which passed the 'Red Flag' law; this prohibited any 'street locomotive' going faster than four m.p.h. on the open road and two m.p.h. in built-up areas. All vehicles had to be preceded by a man on foot, carrying a red flag to warn of its approach. The abolition of this law in 1896 encouraged the development of Britain's motor industry.

Another German engineer, Rudolf Diesel, swiftly followed the

trend with his modified version which made a significant contribution to the development of both transport and industry. Diesel engines were soon being used for generating electricity, and as the motive power for ships, buses and lorries.

The harnessing of electricity began to change people's lives. Edison invented the incandescent electric light bulb. Marconi sent his wireless messages over ever-increasing distances, and after the telegraph came the telephone. In 1881 the first electric tramway line, built by Werner von Siemens, ran through a Berlin suburb. Nine years later, the first electric underground railway – London's 'tube' – was opened. Sir Charles Parsons, a British engineer, invented the first effective steam turbine which he demonstrated dramatically at the 1897 Queen's Jubilee Naval Review at Spithead. With the might of the Royal Navy, in review order, stretching as far as the eye could see, the sudden appearance of Sir Charles at the helm of a tiny turbine-driven boat weaving in and out of the Fleet at the astonishing speed of 35 knots, must have brought Admirals to the verge of apoplexy. The exercise nevertheless achieved its purpose. Within a decade or so most of the world's ocean-going ships were powered by the steam turbine.

The last of the great developments in transport was aviation. While the first giant dirigible airship, named after its German inventor Count von Zeppelin, created a sensation when it ascended in 1900, the first flight of heavier-than-air machines took place at Kitty Hawk in North Carolina on 17th December, 1903. On that cold Winter's morning, the bicycle-shop owners, Wilbur and Orville Wright, with their home-made aeroplane, achieved the first motorised flight, covering a distance of 120 feet in 12 seconds.

25th July, 1909, however, was the eventful day for England. Louis Blériot, a French lamp manufacturer and amateur flyer, set his sights on the £1,000 offered by the *Daily Mail* for the first Channel crossing by aeroplane. 'Why does not Lord Northcliffe offer a prize for a flight to the Moon?' asked a rival paper. 'It would be just as likely to be won.' Blériot hobbled on crutches to his machine – having burnt his legs in his last air accident – and landed half an hour later in the North Fall meadow near Dover. The only witness was a policeman, who ran to the nearest telephone and informed his station: 'That flying man is here!'

Blériot was given a royal reception in London. The press, however, with the exception of the *Daily Mail*, reflected the public mood of apprehension. 'England has ceased to be an Island', was the consensus. The leader in the *Daily News* of 26th July, in more prophetic vein, said:

> A rather sinister significance will, no doubt, be found in the presence of our great Fleet at Dover just at the very moment when for the first time, a flying man passed over that 'silver

streak' and flitted far above the mast of the greatest battleship. Of course, such an event suggests all manner of fresh dangers as well as fresh advantages for the future, and it is one more evidence of some absurdity inherent in our civilisation that our first thought about aeroplanes would be their possible use in war.

But it was left to Lord Haldane, Secretary of State for War, to make one of the classic foot-in-mouth statements of all time, when he told a group of aviation pioneers that 'the War Office could see no use in encouraging aviation, which would be quite useless for war purposes'. A few months later, however, he commissioned the popular flyer and aeroplane designer, Samuel Cody, born in Texas but a naturalised Briton, to submit plans for the first British military aeroplane.

By this time the *entente cordiale* had come back into vogue on both sides of the Channel, and the Tunnel was once again in the news. In fact it was in 1906 that the nickname 'Chunnel' appeared for the first time in the English press. In March the *Daily Graphic* published a railway map of the future, showing how travellers from England might be able to go all the way to New York by rail – via the Channel Tunnel, across Europe and Asia, and through another 15-mile tunnel linking Siberia and Alaska under the Bering Straits.

It was not only the *entente cordiale* which was responsible for the revival of public interest in the Tunnel; another French enthusiast, like Thomé de Gamond before him, was stirring the embers of the project. In 1875 Albert-Henri Sartiaux had been appointed engineer with the *Chemin de Fer du Nord*, where he introduced within the course of three decades a whole series of innovations. He had become the most ardent apostle of the Tunnel and was to dominate the scene until the end of the First World War.

As General Manager of the *Chemin de Fer du Nord* and Government Engineer for Roads and Bridges, Sartiaux produced, in collaboration with the French Channel Tunnel Company, a twentieth-century plan which showed a number of improvements on the 1880 scheme. There were to be two single-track tunnels of 18 feet 5 inches in diameter, connected by cross-passages every 100 yards, and a separate drainage tunnel. Traction would be by electric locomotives – the change-over from steam engines on the French side taking place at Wissant. The trains themselves, Sartiaux argued, would ventilate the Tunnel sufficiently, but he also proposed supplementary ventilation. As electric traction would permit sharper curves, he suggested a new route eliminating the wide semi-circular loop near Calais, with a saving of about six miles of approach line.

Albert-Henri Sartiaux –
appointed engineer for the
Chemin de Feu du Nord *in*
1875. An ardent apostle of
the Tunnel, he was to
dominate the scene for three
decades.

Sartiaux's scheme for a
viaduct out over the sea
near Wissant.

Meanwhile, on the English side, he suggested there should be an international station plus extensive sidings just outside Dover operated by the South-Eastern and Chatham Railway – the result of the merger in 1899 of the South-Eastern with its old rival, the London, Chatham and Dover line.

Perhaps the most original feature of his scheme was an ingenious idea to soothe the nerves of Britain's militarists. On approaching the French Tunnel entrance, the trains would emerge from the cliffs north of Wissant onto a 900-yard-long semi-circular viaduct out over the sea, before entering a portal in the cliffs towards Sangatte and the tunnel proper. Only a short detour, to be sure – but that viaduct would save the island race! At the hint of a threat, the Royal Navy could destroy the viaduct, and no more trains would be able to enter the Tunnel from the French side. A similar viaduct at Shakespeare Cliff could also be built if the French General Staff developed apprehensions about an invasion from England. With this in mind Sartiaux published his plans in 1906 in the form of articles in the *Revue générale des chemins* and the *Revue politique et parlementaire*, as well as explaining his viaduct scheme in a paper entitled 'The Channel Tunnel, from a Military Point of View'.

Like de Gamond before him, Sartiaux sought the co-operation of British engineers, and teamed up with the eminent railway engineer and tunnel expert, Sir Francis Fox, of the Westminster firm of Sir Douglas Fox and Partners[18] who had designed the scheme for a Channel Train Ferry in 1905. The Partnership, now associated with both Albert Sartiaux and the Channel Tunnel Company, were commissioned by the latter to prepare a comprehensive report as the basis for a new Bill which was put before Parliament in late 1906.

After reviewing the geological and technical aspects, the report dealt with the latest developments in electrical traction and lighting: ventilation in a railway tunnel would be no problem, fire prevention easy if no flammable materials were used in construction, or in the building of the rolling-stock. Electric locomotives would be 'armoured' against fire and each

[18] The Partnership exists to this day, as Freeman Fox & Partners, with premises still in Westminster, continuing in the highest traditions of British civil engineering.

carriage would 'carry its own store of light'. Summing up the engineering questions, the report concluded:

> We agree with M. Sartiaux and Mr. Brady in the opinion that the enterprise is one that can be carried out with certainty, and at comparatively moderate cost, the geological and other conditions being of an exceptionally favourable character for the construction of a submarine Tunnel.

The estimated overall cost of the British half of the undertaking was £8 million.

In December of that year the new Bill came before the Liberal Government of Campbell-Bannerman. There seemed to be a great deal of popular support – but to no avail. The echelons of the Military and the Admiralty closed ranks, and the hoary old arguments were revived.

Again, the French failed to comprehend British thinking. In a speech to the influential *Société du Nord de la France* at Lille in January 1907, Sartiaux painted a very serious scenario which would considerably exercise the minds of the military objectors to the Tunnel.

The recently appointed German Chief of Staff, Moltke, had advised the British not to build it. 'I suppose', commented Sartiaux, 'that the famous Moltke is not only a great soldier but also a good psychologist who fears that the Tunnel would have the effect of linking England and France by strong new ties. He knows well that in case of war his own country would not profit by the Tunnel.'

On 5th August, 1913, a deputation of 15 MPs representing the 'House of Commons Channel Tunnel Committee' was received by Prime Minister Asquith. Their spokesman, the Conservative MP

Early 20th century schematic of the geology as it was believed to be at that time. The model includes Sartiaux's scheme for a viaduct near Wissant.

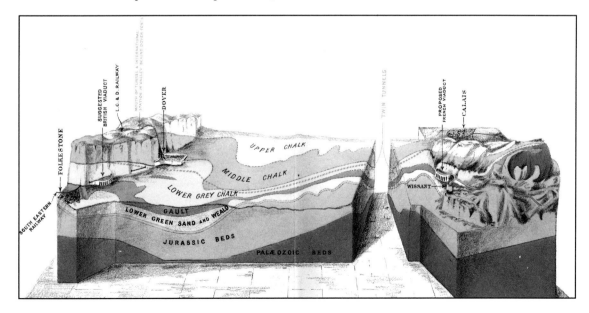

Arthur Fell, handed him a 'memorial', signed by 88 MPs whom Fell had recruited to the Tunnel Committee in June 1913. Introducing the deputation Fell told Asquith:

> This Committee was the result of a spontaneous movement from the back benches, and it has not been engineered in any way by any of the railway companies or by the old Tunnel Company; nor are we concerned with any outside interest in any shape or form. Our object is solely to secure, if possible, that this Tunnel should be constructed . . .
>
> The Committee contains many members, including officers of the army, who were formerly opposed to the tunnel, and who now support it . . . They consider that the question of the food supply of this country in time of war is much more urgent than it was thirty years ago . . . In fact, I have looked through the report of the evidence which was taken then . . . the question of food supply was never once raised . . . We consider . . . that the development of aerial navigation has altered our position as an island, and no man can say what will be the ultimate effect of this. Above all, we consider that our friendship with France, which has stood the test of 98 years under varying conditions, is assured, and that the construction of the tunnel will still further cement this friendship.
>
> What is the estimated cost? [enquired Asquith]. The estimate of the engineers for the double Tunnel, I believe, is about £16 million [replied Fell . . . The Tunnel] would be working with the aid of the Northern Railway of France, which is . . . perhaps the most successful railway company in the world . . .

Asquith replied cautiously, apparently still very much influenced by the attitude of 'our best expert authorities [the Committee of Imperial Defence] who were now reviewing the best interests of the country in the matter':

> Their considerations would be approached with unbiased minds . . . with the single desire to secure on the one hand the absolute strategical safety of the country, and on the other hand the largest, freest possible outlet for trade and inlet for food supplies and raw material, on which the industries of the country so largely depend.

In December 1913 the Channel Tunnel Company published a booklet reviewing the public discussion of the scheme, which was now in full swing. It included details of the plans, past and present arguments, press comments and experts' opinions. Lord Wolseley's scathing attack and Gladstone's reasoned defence of the Tunnel idea were reprinted in full; as was Sartiaux's spirited article in the *Revue des Deux Mondes* of October 1913, which retold the entire history right up to the House of Commons' deputation to Asquith, including the prophetic paragraph:

> If circumstances made it necessary for Great Britain to send, as was the case 100 years ago, a military expedition to the Continent, either to protect some threatened independent nation or to help an ally to maintain the European equilibrium, this expeditionary force could be transported without any risk and without taking up a part of the navy to protect the transports.

Those 'circumstances' were to arise only a few months after the publication of Sartiaux's article.

In that pre-war year, British public opinion seemed to swing in favour of the Tunnel. For a number of months, the *Daily Graphic* campaigned for it; the Liberal *Daily Chronicle* began its second campaign in the Autumn. Even the insular *Daily Express* said that 'the whole question ought to be reconsidered in the light of new facts'.

Meanwhile, the War Office, the Admiralty and the Board of Trade were busy complying with Asquith's request for fresh reports on the problem. These were completed late in 1913 and submitted to the Committee of Imperial Defence which would have the final word. Six months later, on 15th July, 1914, the Committee advised the Government not to proceed.

At the time there was no public announcement, for less than three weeks later War broke out. Before the month of August was over, the British Expeditionary Force embarked for France.

THE FIRST WORLD WAR

Almost at once the need for the Tunnel was felt acutely. In the early stages of the conflict 700 guns had to be rushed from England to France by an improvised ferry service. They arrived almost too late to save the allied front, and many on both sides of the Channel must have regretted, on this and other occasions, that there was no safe and swift connection between the two countries for the movement of troops, guns, ammunition, vehicles and supplies, and a quick comfortable route to bring back the wounded.

Another powerful argument for the Tunnel was that submarine warfare was making surface ships more vulnerable. During the early years of the War German U-boats attacked only warships; but in February 1917, in a desperate attempt to break the stranglehold of the allied blockade, Germany retaliated with 'unrestricted U-boat warfare'. Included in the average 600,000 tons of shipping sunk per month during the remainder of the War there was a high proportion of neutral tonnage. The allied fleets managed to keep the Channel open, but it was never safe from U-boat attacks.

What had been an inconvenience before the War – the transhipment of people and goods on and off boats and trains –

became a hazardous procedure involving dangerous delays. For security reasons, Channel crossings were usually made not by the shortest route but via French ports well behind the front line, such as Le Havre, which meant an additional loss of precious hours.

Contrary to what one might have expected, the Tunnel commanded considerable attention when the need for an invulnerable permanent link became increasingly felt. In July 1915, Arthur Fell, MP, tabled a motion declaring that the War had 'demonstrated the great advantages' which would have accrued to the Allies if a Tunnel already existed, and demanding the immediate start of preparations so that the work could begin as soon as the War was over. It was late in October 1916, that a deputation from the Tunnel Committee of the House, which now numbered 160 members, went to see Asquith, who was still Prime Minister. It urged the Government 'to take a broad view of the question, and to weigh the advantages which will accrue to this country from direct communication with the Continent – against any possible risk of danger in the event of a future war with France'. Asquith replied by reminding the deputation of the adverse decision of the Committee of Imperial Defence in 1914, though he had to admit that the War had changed conditions a great deal. He again promised to instruct the War Committee and the Committee of Imperial Defence that they should 'express a judgement upon the scheme without prepossession or prejudice'.

Artist's impression of a traffic jam in the Channel Tunnel.

In December 1916 Lloyd George became Prime Minister of the Coalition Government, but Bonar Law stayed on as Chancellor of the Exchequer, and it was from him that the answer came to the Commons, first in April and then again in August 1917: 'It is not practicable to proceed further with the matter during the continuance of the war.'

The most important publication on the Tunnel during those years was a 9,000-word article by Albert Sartiaux, 'The Channel Tunnel – During and after the War', which appeared in the *Revue des Deux*

Mondes of September 1917. It was also published, in English, by the Channel Tunnel Company in London. It was bitter and hopeful at the same time; reasoned, yet full of Gallic anger at the waste and suffering that might never have happened if there had been a Tunnel in 1914:

> 'Ships, ships, ships!' exclaimed Lloyd George recently in one of his eloquent speeches. Never before has the merchant service gone through such a crisis as today; never before has there been such a shortage of men as that caused by the war. It is estimated that more than a third, some say even one half of the merchant fleets of all trading nations, which had a total of about 48 million tons in 1913, was lost or put out of action by military operations. Mr. Henderson, formerly a member of the British War Council, estimated in the spring of 1917 that the number of killed in all belligerent nations had reached the appalling figure of 18 million, and that the total number of wounded and killed exceeded the population of the United Kingdom . . .
>
> The consequences of the war are growing increasingly serious – the scarcity of essential commodities such as wheat, coal, meat, petrol, iron, cotton, wool, leather, paper, wood, etc.; the interruption of the exchange of goods which are dependent on exchange as the production centres are often too far away from the centres of consumption; ports and docks congested, the Allies – or at least the English – forced to close their ports to anything that is not vitally important for the war, railway traffic much reduced, personal effort paralysed, the cost of living and labour rising, friction between employers and employees increasing – and above all a distinct danger caused by the exhaustion of the armies in the field . . .
>
> This condition will no doubt be aggravated until the end of the war, and is bound to continue beyond it for several years. After this terrible period of world-wide wear and tear – for there are virtually no neutral countries – the need for effective and multiple means of transport will be more felt than ever before, in order to assure the utilisation of all the resources of the defenders of civilisation for the task of reconstruction, and to enable the Allies to win also the commercial way with which our enemies are ready to follow up the military one . . .
>
> We have, however, in our hands a very powerful weapon which is hardly recognised, if only our British Allies will use it, and which will not only cure many of our troubles when peace has come but will further cement our *entente*. That weapon is the Channel Tunnel.

W. HEATH ROBINSON

Contemporary cartoonist's impression of paddling through a flooded Tunnel.

Sartiaux then went on to calculate the cost which – due to the absence of the Tunnel – had been borne by the Allies, particularly Britain, in the first three War years:

> It has been estimated that the number of journeys made between England and France in both directions since the beginning of the war has been no less than twenty million, for in addition to the transport of troops there has been the continual coming and going of soldiers on leave, the transport of wounded men home and the returning of many of them to the front, and the journeys of all kinds of non-combatants on way duty, with the result that one individual may have crossed the Channel as often as fifty times in one year.
>
> As to the transport of war materials, about 8 million wagons are reckoned to have served the British Army in the French Northern Railway region since the beginning of the war, most of them on the longer route from and to Le Havre, Rouen and Amiens instead of the short one which the Tunnel would have provided . . . When we realise that the transport of one person or piece of equipment from an inland point in England to an inland point in France necessitates no fewer than six operations of embarkation and disembarkation, we must assume that millions of man-hours have been wasted because the Tunnel does not exist.
>
> Then there are the wounded. Is there a family in England with relatives at the front, which would not have been glad to know that there is a Tunnel? . . .

The Tunnel, argued Sartiaux, could help to shift the economic balance of the European countries decisively in the post-war era:

> It is probable that the Tunnel would indirectly make Britain again the economic centre of Europe, and perhaps of the world. In the west, she will be the advance guard of the Continent with the New World; in the east, she will be the centre of attraction for the trade with Europe and the western hemisphere . . . Within, say, fifteen years, when France's devastated provinces have risen from their ashes, Britain and Belgium will form with us the industrial centre of Europe. It was this prospect that Prince Frederick Charles of Prussia feared when he said to the Prince of Wales, afterwards King Edward VII, that the Tunnel would never be built because Germany opposed anything that would destroy her economy.

Finally, Sartiaux appealed to the community spirit of the Celtic nations – with a curious disregard, however, for the Anglo-Saxon element in the British Isles:

> Above all, France and Italy will be united with England – the three sister nations of Western Europe . . . 'The Celtic people of Britain are similar to those of our own country,' wrote Tacitus long ago in his biography of Agricola, Governor of Britannia . . . The development of the English nation has been the same as that of all the Latin races, following the lines of French and Continental culture. Shakespeare belonged to the century of Michelangelo, Cervantes, and Ronsard. In politics, France and England have been the pioneers of freedom . . .
>
> I think that the Channel Tunnel will, by completing the system of the great tunnels under the Alps, make of our three countries (France, Britain, Italy), one single fortress, the impregnable citadel of Liberty.

With America's entry into the War in 1917, the Tunnel began to excite the minds of engineers and financiers. A New York engineer, John K. Hencken, submitted a fanciful plan for its construction – during the War! – to the British and French Governments; he also wrote to Gustave Eiffel, creator of the Eiffel Tower and then in his mid-eighties, asking for his support. Eiffel became interested and entered into a lengthy correspondence with Sartiaux on the American's plan to build the Tunnel in no more than thirty-five days from the English coast only. Using eight machines he had designed, four tunnels would be dug simultaneously at the breathtaking speed of nearly 100 feet per hour. Two would be road tunnels and these would be completed first to allow supplies from Britain to reach the front; the other two would be railway tunnels.

This extraordinary proposal received a great deal of attention from the British newspapers, but Gustave Eiffel was not quite sure if Hencken's figures were right. Even with his wonderful machine, assuming it came up to expectations, the job would take about three and a half months, 'which is slightly less fantastic', as he wrote to Sartiaux. Needless to say the plan came to nothing.

THE MAN BEHIND THE SCENES

To Tunnel partisans on both sides of the Channel, the years after the First World War seemed pregnant with opportunity. Anglo/French concord had been forged in the fire of battle; generals had pronounced upon the vital importance of the Tunnel had it existed; and capitalists in Europe and America were offering to finance it. Yet one disappointment followed another.

It seems incredible, but there was one man to blame. Sir Maurice Hankey was the man behind the scenes of this tragi-comedy who confessed that he had done his damnedest to impede the scheme at the highest level wherever and whenever he could. Sir Maurice Hankey, first Baron, born in 1877, served in the Royal Marine artillery at the turn of the century before transfer to naval intelligence. He became Secretary of the Committee of Imperial Defence for no fewer than twenty-six years – from 1912 to 1938. It was in this capacity that he was able decisively to influence the fate of the Tunnel on several occasions.

In volume II of his biography[19] Captain Stephen Roskill quotes Sir Maurice's diaries and other writings which show the man's lifelong fear and deep-rooted loathing of the Tunnel idea. On 16th November, 1919, Hankey wrote in his diary:

> We had three meetings this week, as well as a great number of conferences. Both Lloyd George and I were very exhausted at the end . . . Rather a scandalous circumstance arose in connection with the Channel Tunnel. A Parliamentary deputation had been promised an interview on the subject, but as it was on the Agenda, the P.M. brought it up at the fag end of an exhausting and exceptionally long Cabinet meeting on Ireland, just as everyone was getting up to go to the Guildhall to the lunch for Poincaré. Although this question has at least 14 times been turned down by Parliament, and has been rejected six times by Committees after prolonged enquiry, including the Committee of Imperial Defence in 1907 and 1914, the P.M. thought fit, without any discussion, and without hearing the views of a single expert, to take opinions. To my surprise the Cabinet were . . . *almost unanimously in favour*[20], Balfour alone saying that his opinion was shaken by a letter I had written to him. This, and a few rather hedging utterances by other Ministers about agreeing, provided there was no risk to national safety, enabled me to record a very feeble support in the official Conclusion. The P.M. and Bonar Law both thought they had secured complete agreement, but I saw them both before the deputation on the following day and convinced them that it would be unsafe to agree without much fuller inquiry into the project. Mr. Balfour, Sir Eric Geddes, and Dr. Addison all came to me to say they hoped I had not recorded a decision in favour of the Channel Tunnel. *What power lies in the draftsman's hands! I could easily, had I been a Channel Tunnel man, have rushed the situation*, recorded a decision, and induced the P.M. to sell the fact to the deputation. In fact he told me afterwards that I only had induced him not to do so. *As matters stand I may be able to block the whole thing . . . I will stop at nothing to prevent what I believe to be a danger to this country.* People have forgotten that 21 years ago our Mediterranean fleet was cleared for action against France; that 18 years ago during the Boer War there was a real danger of a Continental coalition against us; that 18 months ago we were

[19] *Hankey: Man of Secrets*, Stephen Roskill, London, 1972.
[20] Author's italics.

seriously discussing the evacuation of the Channel ports. France may become hostile, Germany or a Bolshevist state may get the other side of the Channel, or the League of Nations may become a mere coalition against us. How should we like the Channel Tunnel then? The increased range of modern guns which can fire across the Channel, aircraft, and the possibility of using 'tanks' walking out of barges as intended in the landing on the coast of Flanders have infinitely increased the risk of a raid – always considered possible. And a successful raid on the British end of the Tunnel converts a raid into an invasion . . .

A week later Hankey wrote in his diary about a meeting of the Home Ports Defence Committee, to consider measures for safeguarding the country if the Tunnel were built. Hankey, as Chairman, listed his arguments against the scheme, 'hoping to kill the Tunnel in spite of the bias in favour of it'. On several occasions during the following decade he went into action against the Tunnel; each time 'he resurrected the opinions which he had expressed against it in years gone by', records his biographer. 'There seems to have been no issue except the preservation of Maritime Belligerent Rights by Britain . . . over which he felt more strongly.'

Artist's impression of the proposed Channel crossing options.

As demobilised soldiers began to form long queues outside the Labour Exchanges, a new argument in favour of the Tunnel arose: that it would help to create some of the promised 'jobs for heroes'.

In July 1919, a French ministerial committee confirmed the concession, which had been granted to the French Channel Tunnel Company some forty years earlier. This move at least stirred the French into action, only to reveal the apathy of the British Government. However, Sir Arthur Fell had appealed to Clemenceau to get things under way.

The French Ambassador asked the Foreign Office whether the British Government was now prepared to support the scheme. In September 1919, the Foreign Secretary replied:

I regret that H.M. Government is not yet in a position to form definite views on the question of the proposed Tunnel under the Channel . . . This question is of the utmost importance, and the various departments of H.M. Government who are interested are now occupied in examining from all points of view the complicated problems raised by it.

In no mind to let matters rest, on 12th November Sir Arthur Fell once again headed a deputation from the House of Commons. The Committee's appeal to Lloyd George was short and simple. It urged that, in view of the experience gained in the War, the Government should 'forthwith allow the construction of the tunnel to proceed.'

Sir Arthur emphasised that there were about 300 members in favour of the project and Lloyd George seemed firmly on their side. He confessed frankly that, in his opinion, the Board of Trade had been wrong in 1907. Some Ministers who formerly had a rooted hostility to the whole enterprise changed their minds – subject to further examination by military and naval advisers. 'If the military advice is favourable', said Lloyd George, 'the Cabinet will certainly be prepared to support the scheme on general grounds.'

Four days later, Lloyd George brought the matter up at the very Cabinet meeting whose 'inside story' Hankey recorded in his diary.

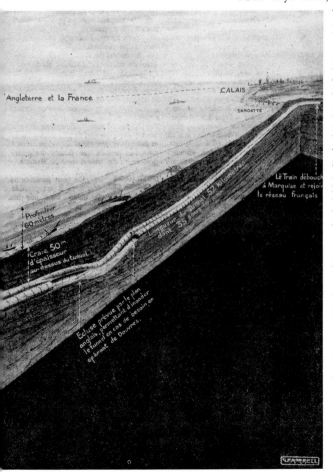

In the House, Sir Arthur Fell kept prodding the Prime Minister almost to the last day before Recess. Could he not arrange for the Government to receive the military and naval experts' advice in time to render possible the deposition of a Bill, asked Sir Arthur, 'so that it might, if passed in 1920, save a whole year's time?' Lloyd George promised to do what he could and, although he did receive the experts' report just before the end of the Session, it was too late for it to be considered.

The Tunnel scheme again ran out of luck in 1920. But was it really only a matter of 'careful examinations' and parliamentary time-tables? The answer came forty-three years later, when a selection of British foreign policy documents was released for publication. The anti-Tunnel pressure group lurking in the dusty corridors of Whitehall was clearly identified. It was the Foreign Office.

In April 1920, the Foreign Office recorded its opinion on the scheme in a memorandum. A few weeks earlier, Marsal, the French Minister of Finance, had told Lord

Derby, the British Ambassador in Paris, that his Government was in favour of starting work on the Tunnel as soon as possible, and asked for an urgent British decision. Surprisingly, the Foreign Office found that the military and naval pundits had dropped their objections. The memorandum asked whether it was desirable to spend £30 to £50 million of Government money on a Tunnel which might, within a few years, have to be destroyed at a moment's notice in an emergency? That did not, however, appear to be the key issue as far as the Foreign Office was concerned; it was the future of Britain's relations with France!

'Our relations with France', declared the memorandum bluntly, 'never have been, are not, and probably never will be sufficiently stable and friendly to justify the construction of a Channel Tunnel.' Until the end of the Napoleonic Wars, the document maintained, France had been England's natural enemy; and even now, in 1920, 'the slightest incident may arouse the resentment and jealousy of the French, and fan the latent embers of suspicion into a flame'. The document then recalled some of the more recent cases of friction: French hostility during the Boer War, the Fashoda incident, the Siam dispute; and it concluded, with astonishing firmness: 'Nothing can alter the fundamental fact that we are not liked in France, and never will be, except for the advantages which the French people may be able to extract from us.'

The Whitaker Rotary Tunnelling Machine.

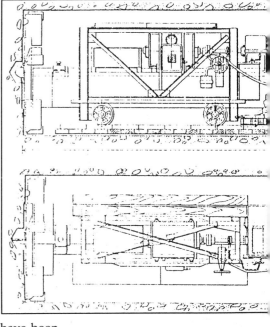

With such scheming in the dark recesses of Whitehall, capable of blocking any Government action in favour of the Tunnel, post-war enthusiasm turned to despair. An entirely new and unforeseen political barrier had surfaced, replacing over a century of military and naval objections to the Tunnel which had evaporated in the light of the salutary lessons learned during the recent conflict. 'Had there been a Tunnel under the Channel in 1914', Marshal Foch told a meeting of the *Cercle Interallié* in Paris in March 1921, 'the war might never have begun.' He later modified that statement to: 'If the English and French had the Tunnel the war could have been shortened by at least two years.'

An army is only as good as its lines of communication and supply. The ability to supply and reinforce the front line overnight during the stagnation of trench warfare would have helped enormously to bring hostilities to an early conclusion. This would have reduced the slaughter, and saved the lives of the many thousands who within 24 hours of being wounded could have been treated in a hospital in England.

Three years later, Sir John French (Lord Ypres) told the

Committee of Imperial Defence that the Tunnel might have affected the strategy of the Allies adversely as they would have had to protect the French exit. Foch repeated his belief, however, and the German military leaders, Admiral von Tirpitz, General Ludendorff, and Chief of Staff von Moltke remained emphatic that the Tunnel would have been a 'threat' to their plans.

Little happened during the immediate post-war years, except for Sir Percy Tempest's[21] interest in a new boring machine designed by Douglas Whitaker of Leicester, which had been effectively used during the First World War. In 1922-23 Sir Percy undertook trial borings with the Whitaker machine in the Lower Chalk near Folkestone, in a cutting above the Martello Tunnel on the Folkestone-to-Dover line. The results were far from encouraging – the cutting-head frequently stuck in the tunnel face and the machine proved difficult to steer. After driving inland some 420 feet of 12-foot-diameter tunnel, work was abandoned. The machine was withdrawn to the mouth of the heading where its rusting remains could be seen half buried in the chalk as if awaiting rescue by some industrial archaeologist[22]. Sir Percy remained confident that an improved version might still be used; he envisaged twenty such machines being built as soon as the Tunnel was approved.

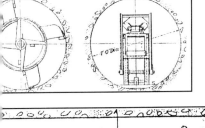

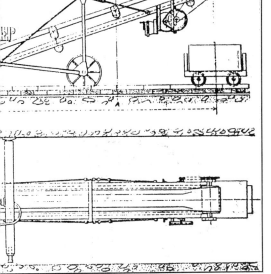

In 1924 Ramsay MacDonald, Britain's Premier since January of that year, showed signs of interest. Sir William Bull, MP, now leader of the pro-Tunnel Committee in the Commons, secured the Prime Minister's promise to review the question without delay and report to Parliament. Naturally, MacDonald turned to the Committee of Imperial Defence – playing straight into the hands of the arch-schemer, Sir Maurice Hankey, who 'at once compiled a long paper recapitulating the whole history of the project', writes Stephen Roskill. He continues:

> Hankey then 'summoned' the former Prime Ministers who had been in power at the time when the Committee of Imperial Defence had deliberated the matter (Balfour, Asquith, Lloyd George and Baldwin) to attend the Committee 'in solemn conclave'; and he obtained an adverse report from the Chiefs of Staff as well. Small wonder that, to quote his own words, the Channel Tunnel was 'flooded out' in 1924 – despite the powerful lobby of businessmen who favoured it.

[21] Chief Engineer to the Channel Tunnel Company.
[22] The Whitaker boring machine was subsequently recovered by Eurotunnel and is a permanent exhibit at the Eurotunnel Exhibition Centre, Folkestone.

This meeting, which Hankey attended to make sure that things went his way, took a mere forty minutes, sufficient to enable MacDonald to conclude that it was the tradition for a British Prime Minister to steer clear of the Tunnel scheme. On 7th July, 1924, he announced in Parliament that the Government would not support it. His was the first Labour Government to oppose the project – although 300 MPs from all parties were in favour. One such was the Member for Epping, Winston Churchill (later knighted). Churchill had been Minister of Munitions in 1917-18, when the 'tools to finish the job' were being produced so efficiently in England, but took such an interminable time to reach the hands of those who needed them. This experience was to make him a lifelong advocate for the Tunnel. He found the attitudes by past Prime Ministers and Governments quite beyond his comprehension, as he demonstrated in his scathing condemnation of the decision taken by the Committee of Imperial Defence when, as Chancellor of the Exchequer, he wrote in the *Sunday Despatch* of 27th July, 1924:

> In forty minutes [at a meeting of the Committee of Imperial Defence] five ex-, or future ex-Prime Ministers, dismissed with an imperial gesture the important and complicated scheme of a Channel tunnel for which the support of four hundred members of Parliament had, it is stated, been obtained.
>
> Four hundred members – Five Prime and ex-Prime Ministers and forty minutes. Quite a record! One spasm of mental concentration enabled these five super-men, who have spent their lives in proving each other incapable and misguided on every other subject, to arrive at a unanimous conclusion.
>
> There is no doubt about their promptitude. The question is was their decision right or wrong? I do not hesitate to say it was wrong.

No further debate on the Tunnel took place until a Committee of the Economic Advisory Council met in late 1929. Its agenda was to consider the various forms of Cross-Channel communication. Under the chairmanship of Sir Edward Peacock, the Committee produced a majority report stating that a Tunnel would be 'of economic advantage to this country', and recommending that it should be constructed and maintained by private enterprise. The report was completed on 28th February, 1930, and submitted to the Government; MacDonald once again side-stepped the issue and sent it for an opinion to the Committee of Imperial Defence, straight into the clutches of Hankey. His biographer writes:

> While the Naval Conference was sitting in London, Hankey was troubled by the resurrection of another of

Sir Winston Churchill, lifelong advocate for the Tunnel.

his most notorious bugbears – namely the proposal for a
Channel Tunnel, which had been dormant since 1924. After
another prolonged enquiry the Home Defence Sub-Committee
(of the Committee of Imperial Defence) produced a report in
May. Then the Chiefs of Staff went into action – no doubt
briefed by Hankey, who produced a paper of his own on the
subject for the Committee of Imperial Defence. On 1st June
Hankey wrote to Robin (his son) that he had killed a scheme
which he disliked almost as much as Freedom of the Seas 'with
the knife of common sense', having employed the weapon of
'lack of economic attractiveness rather than the military
arguments'. MacDonald, he said, had come round to his view
because 'French intransigence at the Naval Conference' had
made him determined not to be 'tied to the leg' of that nation.
The announcement to Parliament was made on the 5 (June),
and the full arguments were published as a Command Paper.

The White Paper was published the next day. Some critics detected
MacDonald's strong apprehensions about the French, which lay
behind these arguments, of which the principal was:

> Having regard to the element of doubt as to the feasibility of
> construction, the weakness of the economic case, the great cost,
> the long period before which the capital expended could
> fructify, and the small amount of employment provided, the
> Government have come to the conclusion that there is no
> justification for a reversal of the policy pursued by successive
> Governments for nearly 50 years in regard to the Channel
> Tunnel.

Even so, when Ernest Thurtle, MP, put a pro-Tunnel motion before
the Commons on 30th June, 1930 – some three weeks after the
publication of the White Paper – 172 Members voted in favour, 179
voted against. Those seven votes may not have been crucial. The
United States, still reeling from the Wall Street crash, lacked
investment capital. Indeed, the economic crisis had become world-
wide and Britain and France were already feeling its impact. Any
Government expenditure on the Tunnel would have been regarded
as unnecessary or even frivolous in the face of growing
unemployment and social unrest.

While Germany's new masters were sabre-rattling, Britain and
France forged closer physical links. A regular car-ferry service had
been started in 1931 between Dover and Calais. In 1936 the Dover-
Dunkirk route was opened for train ferries. The train ferries
necessitated some standardisation of British and Continental rolling-
stock, a prerequisite of a Tunnel rail link.

In France Philippe Fougerolles was now the leading figure on
the Tunnel scene, and it was his construction plan which had been

considered by the 1929-30 British Committee of the Economic
Advisory Council. Working closely with Ellson, Chief Engineer of
the Southern Railway, he proposed to use the same type of boring
machines as had been employed in digging the tunnels of London's
tube extensions since the end of the First World War.

In 1938 a well-known French engineer, André Basdevant,
revived the motorway tunnel idea, which had already been rejected
in 1929 by the British Economic Advisory Council. His proposals –
submitted to the French Chamber of Deputies by Marcel Boucher,
Member for the Vosges – contained some criticisms of the tunnel
route suggested jointly by the British and French geologists and
engineers.

The British reacted strongly, in the person of Baron Emile
d'Erlanger. He wrote to the secretary of the French Tunnel
Company, Georges Bertin, on 11th May, 1939:

> I would like to draw your attention to the fact that M. Basdevant
> is a dangerous man, and that he could very well do great harm
> to the cause of the tunnel. In a lecture to the French Chamber
> of Commerce in London on 2nd May, he declared that the grey
> chalk bed was dangerous as it was not impermeable, which had
> been found during work carried out at Calais where large
> amounts of water had been discovered in that chalk bed. He
> wants, therefore, to dig his road tunnel in the argillaceous clay
> bed. This good gentleman is not aware of the facts . . . if one
> utters such doubts which, to my mind, are completely
> unfounded, one lends much support to the anti-tunnel voices in
> England.

The Baron also mentioned in his letter the man behind the scenes:
'Sir Maurice Hankey – now Lord Hankey – is no longer Secretary to
the Committee of Imperial Defence, which does not stop him, I
think, from exerting the same influence as before on that body.'

Just as in the period immediately before the First World War,
pro-Tunnellers in both countries developed considerable
propaganda activity. Baron d'Erlanger gave a powerful account of
the arguments in the *Daily Telegraph* of 10th July, 1939, shortly
before his death:

> The Tunnel could now be built for £50 million – a cost which
> would be well repaid . . . In the last war aerial warfare was in
> its infancy, but in any future struggle the skies above the
> Channel would be one of the most intensive fields of aerial
> warfare; ships crossing between France and England would be
> attractive targets. On the other hand, troops and material
> passing through the Tunnel would be exposed to no such
> dangers after they had entered and until they debouched, the
> danger to which they would be exposed being no greater than
> that to which they would be exposed while actually embarking

or disembarking from ships . . . If the Tunnel were in being, all the ports of the Atlantic coast of France would be available to British and Allied ships for landing foodstuffs and other materials from overseas which could be carried directly into England by rail. The necessity of having to blockade by submarines or aircraft the Atlantic coast of France, as well as England's coast, would add immensely to the enemies' problems.

Solid, prophetic stuff, but who in Whitehall would listen? Shortly before his death in 1929, Field Marshall Foch had declared: 'The Channel Tunnel would make another war in western Europe an impossible venture.' This *mot* of the old soldier was quoted in a memorandum submitted to General Colson, Chief of Staff of the French Army, by a prominent politician, Le Besnerais, on 2nd July, 1939. He was reporting on the Tunnel scene, concluding with some strategic logistics:

> Recent studies have shown that in case of mobilisation and during the period of hostilities it will be possible to despatch about 120 or even 130 trains per day in each direction through the Tunnel. One infantry division with all its equipment could be transported in approximately 55 trains, so that the capacity of the Tunnel would permit the transport of two divisions per day. The army vehicles would be carried on flat wagons, loaded from platform ramps as it is constantly done in France.

Contemporary cartoonist's impression of Dover port authorities laying the first stone of the Channel Tunnel.

On 9th August, four weeks before the Second World War broke out, *Vu*, the leading French illustrated weekly, published the first instalment of a series of articles, 'whose interest will be obvious to everybody', on the world's great engineering projects. Number one was, of course, the Channel Tunnel. Ten days later, the *Spectator* printed a passionate plea for the scheme by Raoul Dautry, Chairman of the French Channel Tunnel Committee and a Former Director of State Railways, under the heading 'The Case for a Channel Tunnel'. Why, asked Dautry, had the Tunnel not yet been built? Because the plan had never been adequately understood; and he proceeded to expound briefly, from all relevant angles, what he called 'a true civilising agency'. He, too, ended with a glance at the part it would play in the War that was only two weeks away:

> . . . There are those who fear that in case of war the roof of the Tunnel might be

smashed by a bomb. Really! Really! a roof about 50 metres thick, covered with a mass of water constantly in movement another 50 metres deep! This is simply mental aberration.

One most dedicated Frenchman was Georges Bertin, Secretary of the French Tunnel Company. He would spend much of his time in a modest wooden shed in the *Gare du Nord* which bore a large sign saying: *Bureau d'Études de la Société Concessionaire du Chemin de Fer Sous-marin entre la France et l'Angleterre*. Inside there was a large multicoloured relief model of the Channel sea bed, showing its geological layers; hundreds of files with documents from Tunnel history; and mineral samples brought up from the bottom of the sea by Thomé de Gamond – all in little glass bottles labelled in de Gamond's own hand. On a side table solemnly exhibited on red velvet was a chunk of chalk. 'The last piece of chalk excavated in March 1883', Bertin would explain; and then, with a ceremonial gesture, he would produce a small tin box, containing an old and rusting key with a handwritten note attached: *Clef de la Grille du Puits de la Galerie d'Essai du Tunnel Sous la Manche* – the key, he would say philosophically, 'that could have opened England's gateway to the Continent'.

THE SECOND WORLD WAR

Britain declared war on Germany on 3rd September, 1939, and by June 1940 Hitler had become master of the whole of Northern France including the Channel ports, which put a stop to all Tunnel schemes. However, the idea of digging a tunnel from the English side only, as an advance supply route for the opening of a Second Front, was later discussed at Allied Headquarters: a somewhat impractical idea since the French coast was in German hands. Yet the Germans, with Sangatte in their control, might reopen the pilot tunnel, widen it, and drive toward the English coast – a question which the war leaders in London had at least to bear in mind.

Two wartime committees – the Engineering Advisory and the Scientific Advisory Committee – had been set up, and the Chairman of the Joint Committees in May 1941 sent a memorandum to the War Cabinet, warning Britain's leaders of a possible German invasion by Channel Tunnel. That Chairman was none other than Lord Hankey who now had the chance to remount his hobby horse. The idea of the Germans digging an invasion tunnel 'sounds fantastic at first sight', he admitted, 'but I do not think we ought to dismiss the possibility without examination'. So the opinions of a number of experts were sought.

Sir Henry Dale, President of the Royal Society, calculated that the Germans could do the job in sixteen months. Sir Leopold Savile,

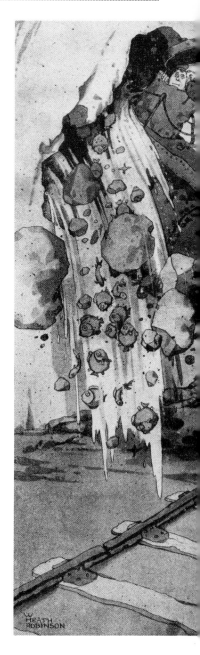

Contemporary cartoonist's impression of a submarine collision with train passengers in Tunnel.

President of the Institution of Civil Engineers, suggested that boring from one end only would take a minimum of twelve years and six months. Other experts theorised that the invasion tunnel could be dug at the working rate of London's Tube – 30 feet per three-shift day – which would mean a total of six years.

Faced with such diverse opinions Britain's leaders could not take any chances, and RAF reconnaissance flights were ordered over the French coast to photograph any evidence of large construction sites, while the Royal Navy was to look out for any discoloration of the sea which would indicate the disposal of Tunnel spoil. No further action was taken for over a year. Lord Hankey was replaced as Chairman of the Joint Committees by the Minister of Education, Richard Austen ('Rab') Butler (later Lord Butler) in March 1942. The War Office and the Admiralty, in July, asked the Joint Committees to send a team of experts to the Dover/Folkestone area with a view to establishing listening posts in the only accessible tunnel, at Abbots Cliff.

Successful experiments in the detection of undersea noises, such as the digging of a tunnel, had been made with quartz piezo-electric listening devices, and Butler sent a secret memorandum to the War Office suggesting that 'routine listening carried out by six men of the Royal Engineers using this apparatus would . . . guard against any risk there may be of surprise by a German Channel Tunnel'. There is no record whether that suggestion was put into effect; as the Abbots Cliff heading runs through the cliffs, not under the sea, the listening posts could not have heard very much. Still, the Navy correspondent of *The News Chronicle*, Vernon Brown, wrote in 1946 that 'a 24-hour vigil was maintained until all danger of invasion had passed'.

Hitler's aborted Operation Sealion has passed into history, but whether the invasion of England by tunnel was ever considered by the German High Command is not known. Documents found after the War gave no indication of such intent. If it had been thought of it would have – without doubt – been regarded as hazardous a venture as had been concluded by the Allied Command.

If the Tunnel had been in place by September 1939, would the drama of Dunkirk have been necessary? Even if allied reverses could not have been checked by reinforcements and supplies via the Tunnel, then at least the British Expeditionary Force would have been assured a dry run home. Would Hitler have tried to turn the British militarists' nightmare into reality, by attempting an invasion of England by sea, air and Tunnel – just as the old French cartoon had visualised? Would the 'Second Front' – D-Day – have been achieved a year or so earlier, shortening the War and saving thousands of lives?

In terms of the Tunnel, the lessons of the First World War

went unheeded and as Hegel put it: 'What history teaches is that neither nations nor governments ever learn anything from it.'

Dunkirk – a dramatic example of the need for a Channel Tunnel.

THE POST-WAR YEARS

Although the offices, archives and most of the shareholders' register had been destroyed during the London Blitz, the 1872 Channel Tunnel Company still had a City address and £11,219 in the bank. Its chairman was Sir Herbert Walker, a director of the Southern Railway which, with 26% of the shares, was the major shareholder. The company also had a most important asset in the continued support of the d'Erlanger family. Leo d'Erlanger was present with Sir Herbert in September 1947 when the company met for precisely three minutes. The secretary, a director and three shareholders all agreed that post-war economic circumstances were hampering the promotion of the Tunnel scheme. Then Sir Herbert rose to suggest a motion which shocked the meeting. 'I propose', he said, 'that we wind up this whole joke.'

He was at once asked to withdraw his motion if indeed he was being serious, and Leo d'Erlanger undertook to brief him with the facts and arguments. Sir Herbert returned a few weeks later a total convert. 'I do not hesitate', he said, 'to propose that we go ahead at full speed.'

Leo d'Erlanger himself had been anything but an enthusiast when, after the death of his uncle, Emile Beaumont, just before the outbreak of World War II, he became head of the family firm of merchant bankers. Outside the City, his early interest was flying, and although family loyalty demanded his support for the Tunnel, he remained something of a sceptic. The War, however, had convinced him, and wherever anything happened on the Tunnel scene, Leo d'Erlanger was there – eloquent, sincere, and a true enthusiast.

Early in 1948, British and French experts and politicians met in Paris to discuss new plans to be laid before their respective parliaments. The spokesman of the British delegation was Commander Christopher Powell, RN, Secretary of the All-Party Channel Tunnel Parliamentary Group. His French colleagues expressed great interest in campaigning for the project. They favoured one tunnel incorporating a two-lane motorway, while the British favoured two rail-only tunnels; but both sides agreed that the cost would be about £50 million.

The climate in Britain was now more favourable. The newly nationalised railways were now on an equal footing with the French State Railways, and both favoured State ownership of the Tunnel. France, too, had its Parliamentary Group, with Guy d'Arvisenet as its secretary. France's Tunnel project had been planned by André Basdevant, whose single-tunnel design had so displeased Baron Emile d'Erlanger before the War.

In March 1949, 200 British MPs of all parties tabled a motion urging the Government to submit the new Tunnel plans not only to France, but to all Western European countries, 'in order to achieve a close political, economic, and cultural union of Britain and Europe, essential to their common well-being and stability'. A rapid, regular and safe physical means of communication was necessary since sea and air transport by themselves could never cope adequately with the traffic of the future. The time was coming, added Mrs. Leah Manning, MP, when the 'little man' would be the 'little man in the car', and it would be the ordinary people of Europe who would reap the benefit of a Channel Tunnel.

The enthusiastic André Basdevant, together with Ernest Thurtle, MP, whose 1930's motion in the House of Commons had been so narrowly defeated, submitted the rail-road Tunnel plans to SHAPE (Supreme Headquarters, Allied Powers, Europe) in 1951. Two years elapsed before SHAPE expressed its support from the military standpoint. The French Channel Tunnel Company, instead of welcoming the decision, was annoyed because it feared that its rights as one of the Tunnel concessionaires were threatened. In 1953 it endeavoured to assert its position, only to incur adverse publicity for the entire scheme. However, Basdevant's plans had not yet been abandoned.

The hopes which the Tunnel partisans had pinned on the British Labour Government never materialised; nor was the 1951 Conservative Government greatly interested. There was, for instance, a rather depressing ministerial answer in 1954 during Question Time in the Commons:

> *Lance Mallalieu (Labour)*[23]: In view of the desirability of demonstrating the solidarity of the peoples of France and Great Britain in peace and in war, would the Minister of Transport withdraw the objections which successive British Governments had raised to the construction of a tunnel under the Channel?
>
> *Hon. Mark Lennox-Boyd (Conservative):* While there are so many useful and necessary transport projects in this country which have more pressing claims on the limited resources available, I am afraid the Channel Tunnel will have to wait.
>
> *Mallalieu:* At least six continental Governments wish to contribute to the scheme. It would be a suitable scheme in this, the fiftieth year of the *entente cordiale.*
>
> *Lennox-Boyd:* While desiring in every way to show our affection for the peoples of Europe, I cannot concede that the old objections have been removed.

Were those strategic objections still haunting Whitehall? If so, they were resolutely exorcised in 1955 by the Rt. Hon. Harold Macmillan, then Minister of Defence. After an earlier fruitless exchange of opinions in the Commons on 2nd February, to the question, 'What was stopping private enterprise from financing the project?'[24] the Ministry of Transport replied that it 'did not favour the project and saw no possibility of others undertaking the work.' Lance Mallalieu persisted and on 16th February asked Harold Macmillan: 'To what extent do strategic objections still prevent the construction of a tunnel under the Channel?' Whereupon the Defence Minister surprised everyone by replying: 'Scarcely at all.'

The next year saw some interesting behind-the-scenes developments. Early in 1956 Leo d'Erlanger made the acquaintance of Paul Leroy-Beaulieu[25], Financial Attaché to the French Embassy in London, and a director of the *Société Concessionaire du Chemin de Fer Sous-marin entre la France et l'Angleterre*[26]. Their family affinities gave added stimulus to their discussions and they came to

The entrance to Beaumont's 1880 Tunnel.

---

[23]Joint Chairman of the All-Party Channel Tunnel Parliamentary Group.

[24] Estimated to cost £75 million.

[25] Grandson of Michel Chevalier, founder of the 1870 French Channel Tunnel Company.

[26] 1870 French Channel Tunnel Company.

the conclusion that the most urgent task was to interest private capital in the venture. Leo d'Erlanger felt that the project needed the impetus which could only come with the support and backing of a company or organisation 'which was above reproach and of international repute'. Several months elapsed before Leroy-Beaulieu asked d'Erlanger to come to Paris as he believed he had found the company which matched those specifications – *La Compagnie Financière de Suez* – the Suez Canal Company. Jacques Georges-Picot, its Director-General, was greatly interested in their proposals. The 99-year-concession to operate the Canal would expire in 1968, and it was unlikely to be renewed by Egypt. As their failure had been based on the transport feat of nineteenth century engineering, it was apt that they associate themselves with what would be the greatest transport project of the twentieth century.

Then, that very Summer, in response to the withdrawal of British and American finance from the Aswan High Dam, Egypt's President Nasser nationalised the Suez Canal, its property and assets.

Things now looked extremely bleak. It was inconceivable that the Suez Canal Company could maintain its interest in the Tunnel. However, as a banker and a realist, d'Erlanger was used to quirks of fortune and had set his mind to look elsewhere. Later that year, the Suez crisis receded and Georges-Picot informed the Channel Tunnel Company that the Suez Canal Company, despite the loss of its Egyptian assets, was still interested in the Tunnel, and prepared to take its association a stage further[27].

THE CHANNEL TUNNEL STUDY GROUP

In the Autumn of 1956 chance again intervened. Two sisters, on their return to America from a European holiday, suffered severely from sea-sickness whilst crossing the Channel. This in itself was nothing out of the ordinary, but the sisters were heiresses to the large Belgian electronics concern, Schlumberger, and recounted their unpleasant experience to their husbands, Arnaud de Vitry d'Avaucourt, an engineer, and Frank Davidson, an international lawyer, asking what had happened to the idea for a Channel Tunnel. De Vitry and Davidson undertook some research, and were quick to grasp its financial potential. They co-opted Cyril Means, an arbitration director of the New York Stock Exchange, to travel to Europe to explore the intent and purpose of the various groupings of Tunnel partisans.

---

[27] In 1958 the United Arab Republic offered compensation to the Suez Canal Company's shareholders, to be paid in instalments over a period of five years.

Early in 1957, Cyril Means returned to America, and his enthusiastic report led to the foundation, by Davidson and de Vitry, of Technical Studies Incorporated, New York, with the aim of backing a preliminary investigation of the Tunnel project and, if it was feasible, to promote it financially. Technical Studies commissioned a London firm of consulting engineers, Brian Colquhoun & Partners, to undertake a survey[28] of the principal aspects. Their report, 'Project of a Channel Tunnel: Recommendations for Research and Investigations' was submitted in April 1957.

During the month of April, Georges-Picot announced that the Suez Canal Company was now ready to participate and that an organisation was to be set up to carry out a preliminary site investigation and geological survey of the Channel sea bed as well as a study of the general, technical and economic problems.

For a time it looked as though Technical Studies would be left out of the running. There was no great enthusiasm among the new consortium to include the Americans – at least not at this stage. Brian Colquhoun & Partners, further complicated the issue by giving a presentation to a group of MPs in the House of Commons. This brash intervention gave rise to some consternation and a stiffening of the resolve that the Americans had no part to play: it was, after all, 'a purely Anglo-French affair'. Leo d'Erlanger, who foresaw the problems and dangers of two factions competing towards the same goal, intervened and announced at the Annual General Meeting of the Channel Tunnel Company in May 1957 that there was still a role for the Americans, who had the backing of two of the leading US private banks, Morgan Stanley and Dillon Read. The peril of any conflict of interest was avoided, particularly as Technical Studies emphasised that it had no intention of imposing any form of 'dollar domination'.

The outcome was the birth in Paris of a new syndicate on 16th July, 1957 – the Channel Tunnel Study Group – comprising the 1872 Channel Tunnel Company; its French counterpart, the 1875 *Société Concessionaire du Chemin de Fer Sous-marin entre la France et l'Angleterre; La Compagnie Financière de Suez;* and Technical Studies Inc. of New York.

Each participant of the Channel Tunnel Study Group subscribed to the Group's capital of £225,000, with Technical Studies holding amounting to only 10%. Although they were no more than four in number, their ramifications were widespread. The

---

[28] Technical Studies Inc. commissioned three further studies: 'Preliminary Engineering Analysis' (by Howard King, of Mason & Hanger-Silas Mason Co. Inc.); 'Proposed English Channel Tunnel Traffic and Revenue Studies' (Walter P. Hedden); *'Tunnel sous la Manche'* (André Basdevant, Jacques Menager, Marcel Chardel, André Guerin).

principal shareholder of the Channel Tunnel Company was the British Railways Board which had 'inherited' the old Southern Railway's interest of 26% when the railways were nationalised. The main shareholder of the French Channel Tunnel Company was the *Société Nationale des Chemins de Fer* (SNCF), which, together with the Paris branch of the International Road Federation, held a 50% interest; and one of the major shareholders in the Suez Canal Company was, of course, still the British Government, which owned some 33% – the remains of Disraeli's financial coup of 1875. The Study Group was backed by a formidable syndicate of international bankers[29], and had as Joint Presidents René Massigli, the former French Ambassador to London, and Sir Ivone Kirkpatrick, the distinguished diplomat who had been British High Commissioner in Germany after the War.

One of the most respected of the Study Group's distinguished array of experts was Louis Armand, Chairman of EURATOM, President of the International Railway Union, and Honorary Chairman of SNCF, probably best known as the man who had, within ten years, turned the post-war wreck of the French Railways into Western Europe's most efficient public transport system.

Early Channel Tunnel Study Group Schematic.

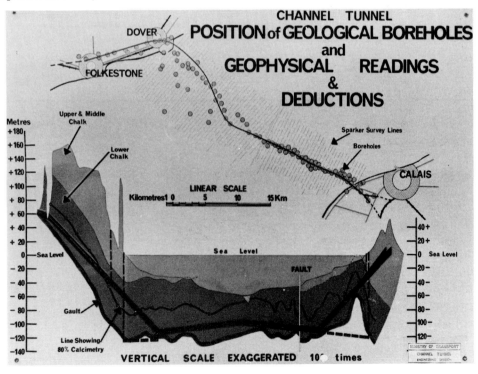

---

29 Morgan Grenfell & Co. Ltd.; Phillip Hill, Higginson, Erlanger Ltd; *de Rothschild Frères; Banques de l'Union Parisienne; Compagnie Financière de Suez*; Morgan Stanley & Co.; and Dillon, Read & Co., Inc.

A team of international experts drawn from the Economist Intelligence Unit Ltd., London, the *Société d'Études Techniques et Économiques,* Paris, and de Leuw, Cather & Co., Chicago, produced the 'Channel Tunnel traffic and Revenue Report'. René Malcor (*Ingénieur en Chef des Ponts et Chaussées*) and Harold Harding (later knighted), one of the world's leading consulting engineers on tunnelling, prepared, together with the distinguished geologist, Professor J. M. Bruckshaw of Imperial College, London, and in conjunction with several engineering consultancies[30], an engineering and geological survey.

Meanwhile, on the London Stock Exchange, the Channel Tunnel Company's four-shilling shares, trading for years at around four pence, suddenly soared to fifty-seven shillings.

The Study Group's geological and geophysical site investigation began in 1958. A seismographic survey was undertaken – employing Sparker, the new sounding apparatus, a continuous seismic profiler which had been conceived by Hersey & Knott and developed at the Woods Hole Oceanographic Institution. Producing low-frequency, high-energy impulses, the Sparker achieved deep penetration of the rock strata down to the base of the Lower Chalk, producing a detailed map of the strata configuration. Ten core bores were undertaken, four inches in diameter, to depths of 225 feet and bottom-sampling and similar methods were employed to ascertain the dimensions of the Lower Chalk. The purpose of the survey was to check whether there were any fissures or faults which might seriously impede the construction; and to discover whether the parameters of the Lower Chalk could contain the dimensions of a bored tunnel.

The 1882 Tunnel workings at Sangatte and Abbots Cliff were inspected and it was found that, while the Germans in World War II had taken the precaution of flooding the French Tunnel, neither Tunnel revealed any major seepage of water. This provided important evidence as to the nature and texture of the Lower Chalk stratum. Technical Studies Inc., which funded the cost of the survey and investigation, now became a full partner in the Group with a 25% interest.

WHAT KIND OF LINK?

The Channel Tunnel Study Group did not confine itself to a tunnel: it commissioned independent studies which included bridges, and

---

[30] In Britain: Sir William Halcrow & Partners; Livesey & Henderson; and Rendel, Palmer & Tritton; in France: the *Société Générale d'Exploitations Industrielles* (SOGEI); Fougerolles; *Grands Travaux de Marseilles*; and Soltranche; in America: Parsons, Brinckerhoff, Quade & Douglas; the Bechtel Corporation; Brown & Roote Inc.; and Morrison Knudsen & Co.

The Abbots Cliff Tunnel, which, although unlined, proved for nearly 100 years the integrity of the Lower Chalk.

bridge-tunnel combinations as well as road, rail, and road-cum-rail tunnels. It also considered a cross-Channel dam, along the top of which would run a motorway punctuated by swing bridges to permit the passage of ships. Also considered was the hovercraft, though at that time the development of this unique form of transport was still in its infancy. These studies were concluded in 1959, and the Group was nearing its moment of decision.

Road tunnels presented grave technical problems of ventilation. Concentrations of carbon monoxide beyond acceptable limits would be unavoidable at peak traffic periods of, say, 750 cars per hour. There was also the human problem: could the average motorist drive safely through some 32 miles of tunnel at 35 miles per hour? The Automobile Association later stated:

> A railway tunnel is to be preferred because we think it would not be possible for anybody to drive that distance in a confined space. When you consider the effect of tunnel-travelling on your nerves, it would not be the best plan for motorists. We feel that a rail tunnel has more merit.

A speed limit of 35 m.p.h. might seem excessively cautious, but a motorway-style pile-up – all too familiar today – in the confined space of an undersea tunnel would be horrific.

A road-rail tunnel would bring with it the ventilation problems of a road-only tunnel and would cost more, as the tunnel would have to be larger to accommodate both methods of transportation; and, in the economists' view, to marry alternative, competing transport systems would be inefficient and uneconomic.

A Channel bridge, almost as old as the Tunnel scheme and with considerable popular appeal, had much to recommend it: technically feasible, it could combine both road and rail transport, and could cope with a higher density of peak-hour traffic than any other kind of link. It was also twice as expensive to build, and had a high annual maintenance cost, because of the corrosive action of the sea and weather. The overriding and insurmountable difficulty was that the Channel is the busiest international waterway in the world, with up to 1,000 vessels a day plying their trade. To build a bridge, as envisaged at that time, with 168 supports, would create permanent navigational hazards. Before construction could start, permission would have to be sought from some eighty maritime nations who regularly use the international waters of the Channel. To put such a plan to the International Court of Justice at The Hague would tax the minds and line the pockets of maritime lawyers with little hope of ever finding a solution.

As for a bridge-tunnel-bridge scheme, cars and trains would leave shore on a bridge section, descend into a tunnel, and complete the journey having ascended to a second bridge section. The central tunnel would leave an unobstructed sea lane open to shipping. However, the piers supporting the bridge sections would still be dangerous to inshore shipping, and the cost would be infinitely greater than that of a conventional bridge[31].

The Channel Tunnel Study Group, after carefully sifting all the evidence, came to the conclusion that twin railway tunnels would be technically the most acceptable, and financially the most viable means of permanently linking Britain and France. There would be two parallel 21-foot, 4-inch diameter one-way tunnels, joined by cross-passages every 250 yards to a central pilot tunnel, 10 feet, 10 inches in diameter; during construction this pilot tunnel would be driven ahead of the main tunnels in order to check and, if necessary, treat the ground through which the main tunnels would be bored. Four cross-over junctions would connect the two main tunnels and, on completion, the pilot tunnel would become a service gallery providing a safety escape route and maintenance access for the entire Tunnel complex.

There would be two Tunnel terminals, one between Ashford and Westenhanger and the other at Sangatte. The distance between the terminals would be 44 miles, and the Tunnel 32 miles long – with 23 miles some 125 feet below the sea bed. No tunnel is ever completely waterproof, and any seepage would drain down into sumps at the lowest points of the lazy 'W' longitudinal configuration of the Tunnel and pumped out.

Electric locomotives would be powered from an overhead 25 k.v. system. On emerging from the Tunnel, shuttle trains, carrying road vehicles, would travel around a loop of railway track within which would be sited the loading and unloading platforms of the Tunnel terminal. Vehicles would drive off the shuttle trains and departing vehicles would drive on for the return journey. Shuttle trains would comprise double-decked, enclosed coaches with a capacity for about 300 cars per train. Operating on a five-minute headway, the shuttle system would achieve a peak-hour capacity of some 3,600 cars in each direction. Drivers and passengers would remain in their cars –

A later model of the double-decked, enclosed coach on display at the Eurotunnel exhibition centre in the 1980s.

---

[31] In addition to the more or less conventional schemes, there emerged, following the Study Group's submission to the two Governments, a proliferation of highly inventive ideas: they were original, ingenious and bizarre, but all were impractical, and a fuller description of some of them can be found in Appendix I.

no particular hardship as the journey would take about thirty-five minutes. Road transport vehicles would be carried on open low-loading flat wagons.

For the rail passenger, there would be no need to change trains. The national rail networks would connect with the shuttle circuit at each end of the Tunnel and passenger and freight trains would travel straight through to their destinations. The travelling time from London to Paris, or London to Brussels, would be at least halved and the journey times of trains to other Continental destinations would be considerably reduced.

TUNNEL FINANCE

The Study Group report, based on the traffic and engineering studies, estimated the cost of the entire project at £130 million. The cost of the Tunnel would be £80 million, with financial charges and interest during construction accounting for some £14 million, plus £6 million for working capital and expenses. The railway terminals, equipment and rolling-stock to be provided by the two State railways would cost about £30 million. The project, therefore, would require a private investment in the region of £100 million.

The International Tunnel Company to be formed would be responsible for the financing and construction of the project. The equity, some 20% of the total investment, would carry the construction beyond the point of engineering risk. The balance of 80% would be by a bond issue, to be raised as and when required during the estimated five-year period of construction.

The Channel Tunnel Study Group sought the concession that it should be secured against any risk of cost overrun during the construction period. The Study Group considered that this would not be a burden on the Governments as the contractors building the Tunnel would most certainly be required to assume a substantial part of any overrun costs.

The Study Group proposed that a large portion of the required private capital should be raised on the international money market outside Britain and France. Because important institutional investors at that time, particularly in the American market, could legally invest in overseas stocks and shares only if they were backed by the Government of the country of origin, it was proposed that the British and French Governments should, on completion of the Tunnel, lease it from the Study Group on a long-term basis. The two Governments would in turn sub-lease the Tunnel to their respective railways, thus providing overseas investors with requisite Government involvement.

The annual rental for leasing the Tunnel to the two Governments was estimated at about £6 million – the amount required by the Study Group to meet its obligations in respect of bond interests and amortisation. This sum – £3 million from each country – was only to be paid if the annual receipts of the Tunnel fell below that figure, and even then only in proportion to the shortfall. As the first year's receipts were conservatively estimated to be in excess of £13 million, there was little likelihood of the Governments ever being called upon to pay rent.

Furthermore, it was anticipated that 25% of the capital would be raised in Britain, with similar participation by France and the USA, and the remaining 25% contributed by other western European countries. Thus, England and France would enjoy all the advantages and facilities of the Tunnel, three-quarters of which would have been paid for by private investment from other countries.

Following submission to the two Governments on 27th March, 1960, the Channel Tunnel Study Group released its proposals to the media – which condemned them almost unanimously!

Clearly the Study Group, absorbed as it had been in producing a financial plan for both a technically and economically feasible project, had failed to anticipate public reaction to the complex nature of the proposal. Against the background of Britain's economic position, the Channel Tunnel appeared as an extravagance that would siphon off large sums of public money from badly needed domestic requirements, such as houses and hospitals, roads and schools. It was also argued that resources of men and materials would be diverted from these national priorities.

The Study Group had recognised – perhaps belatedly – the wisdom of Abraham Lincoln that 'public sympathy is everything; with it nothing can fail, without it nothing can succeed'.

Some four months after its initial submission, the Group filed a supplementary memorandum containing modifications of its proposals, particularly in respect of finance. British Rail was in severe financial straits, being some £100 million in debt and it was unrealistic to expect it to assume the burden of its share of the £30 million for Tunnel terminals, equipment and rolling-stock. The Group would, therefore, assume financial responsibility for both British and French terminals. Thus, the Group was prepared, unconditionally, to shoulder the total cost of £130 million and, as a demonstration of faith, withdrew its requirement for Government responsibility for overrun costs.

The Study Group followed up with a press conference at Brown's Hotel, in London, on 14th July, 1960. Its spokesman, Leo d'Erlanger, stoically faced a barrage of questions and heckling. Yet this tall, elegant figure held the floor with such courtesy, charm and confidence that he earned the grudging admiration of the 80 journalists who held him at bay. Such was his performance that after more than two hours they paid him a rare tribute – applause.

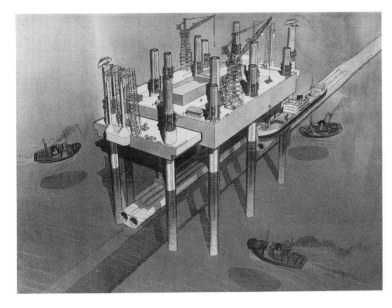

Immersed tube method of construction conceived by Richard Costain Construction Ltd.

The ensuing media coverage, though still far from enthusiastic, offered a cautious appraisal of the possible advantages of the scheme.

The main report submitted to the two Governments in March 1960 included brief details of the two ways of building the Tunnel – the bored and immersed tube methods[32]. There was little to choose between them in terms of time or cost, although the immersed tube had the attractive feature of fixed price construction, backed by a guaranteed bond of completion, and as an alternative introduced an element of competitive tendering.

There was, however, the disadvantage that a deep trench would have to be dredged in the sea bed up to depths of 250 feet to contain the prefabricated tunnel sections. The best-conceived plan for an immersed tube tunnel was produced by a British, French and American engineering consortium, headed in London by Richard Costain (Construction Ltd.). It proposed that the dredging and laying of the prefabricated lengths of tunnel should be undertaken from two vast platforms based on the de Long-type drilling platform, similar to the rigs employed at the time for oil and gas exploration in the North Sea, with the innovation that these platforms would 'walk' across the sea bed.

Slotted into the main deck of the drilling platform would be a second deck, supported by its own set of legs, resting on the sea bed. The 'walking' effect would be achieved by one deck retracting its legs and

[32] The Immersed Tube method was first proposed in 1803 by the engineer Tessier de Mottray and his colleague Franchot.

sliding forward under hydraulic power and, when fully extended, lowering its legs to the sea bed. The second deck would repeat this movement and, by successive actions, the rig would 'walk' to its new site. Two of these vast platforms, working in tandem, would be sufficient to complete construction.

In the past bridge schemes had been repeatedly rejected, but in April 1960, Jules Moch, a former French Minister of the Interior, announced that he had formed a Channel Bridge Study Group (*Société d'Études du pont sur la Manche*) with the support of the *Union Routiér*, an organisation representing French road interests.

Jules Moch's bridge plan[33] appeared feasible enough, although similar plans presented by the same three companies had already been vetoed by the Channel Tunnel Study Group.

His bridge would span the Straits between South Foreland, near Dover, to a point south-west of Sangatte. It would have a central, five-lane highway for motor vehicles, flanked by two railway tracks and two outermost 13-foot-wide tracks for motorcycles and bicycles. The 140 reinforced concrete piers to support the 738-foot-long steel spans, would be prefabricated and floated to their permanent sites, where they would be sunk to depths of up to 165 feet from floating work platforms. Ten longer spans – 1,447 feet – would provide a clearance of 230 feet above high water. The cost of the bridge was estimated at £250 million, considerably more than the Tunnel.

The French public thronged to see the Bridge Study Group's Paris exhibition of a scale-model of the bridge, above a paper sea, with model trains and cars (the latter driving on the left as a concession to the Island Race), and even a model of the liner *ÎLE DE FRANCE* passing underneath it. The French press greeted the scheme with some enthusiasm.

As the Channel Tunnel Study Group admitted, the idea was 'very attractive'. Such a bridge would provide almost unlimited capacity for both road and railway traffic; the crossing would be fast, with no ventilation problems, and diesel locomotives could be used. However, the economics were against it. According to the sponsors' own figures, its profit during the first 12 years, would be less than 3%, making it practically impossible to raise the required private capital.

Meanwhile in Westminster, the game of question and non-committal answer continued:

> I cannot say when any announcement will be made.
> *Ernest Marples, Minister of Transport, 11th April, 1961.*

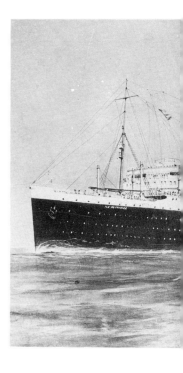

The Île de France.

---

[33] The plan was designed by Dorman Long (Building & Engineering) Ltd. in Britain, *Compagnie Française d'Entreprises* and the Merritt Chapman & Scott Corporation of America.

Progress will be made as best it can.

Lord Chesham, 22nd June, 1961.

I shall not be in a position to make a policy statement until a later stage.

Ernest Marples, 26th July, 1961.

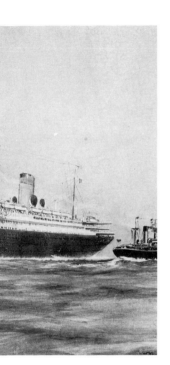

Yet in July, the British Government announced that it would apply for full membership of the Common Market. In August, the Chief of Government Information in Paris revealed that talks about a 'Channel link' would be opened between the two countries and that Jules Moch's bridge scheme not only had financial support but also the endorsement of trade associations such as the *Fédération Nationale de l'Automobile*, the *Industrie du Pétrole, Sidérurgie*, and such industrial giants as *Sud Aviation, Panhoet*, and the *Société des Forges et Ateliers du Creusot*. One most important voice in France, however, that of Louis Armand, remained steadfastly in favour of the Tunnel. His main argument seemed difficult to counter: that the lower cost solution of the Tunnel would mean cheaper rates for carrying people, cars and goods.

The British and French Governments – represented by the Ministers of Transport, Ernest Marples and Marc Jacquet – agreed in November to set up a working group of officials from both countries under the Joint Chairmanship of D. R. Serpell, Deputy Secretary, Ministry of Transport, and J. Ravanel, *Commissaire au Tourisme*. Their brief was to examine and report on the proposal for twin rail tunnels and bridge scheme – both ventures to be privately financed.

In May 1962 Jules Moch gave a press conference in London to promote his proposal, but his bridge engendered little enthusiasm. He claimed, for instance, that the bridge would overcome the 150-year-old military objections to a permanent link:

> The protagonists of the Tunnel claim that in the eventuality of a future warlike operation with Europe, they would flood the Tunnel with water, but if *I* were the enemy I would pump through twenty atmospheres of air, thus clearing the Tunnel of water! I would land paratroops at the British Tunnel portal to establish a bridgehead sector and then use the Tunnel for the passage of my main forces. In similar circumstances with the bridge, I would train the 'big cannon' on the main spans in mid-Channel, and *pouf!* The bridge would all fall down!

An ex-naval captain in the audience commented, rather saltily, that he had a pretty fair idea of the Sea Lords' reaction to any scheme that would 'clutter up the Gut [the Royal Navy's affectionate term for the English Channel] with a mess of bloody ironmongery'. The British press behaved with commendable restraint throughout.

Lord Gladwyn, who had been British Ambassador in Paris two

years earlier, tried to stimulate some enthusiasm in Britain for Jules Moch's bridge. He called a meeting of road transport interests, among them the Road Haulage Association which had always favoured a bridge solution.

At the meeting, Lord Gladwyn proposed a fighting fund of £25,000 for publicity in favour of the bridge which was greeted with some support. On being questioned as to the financial contribution that might be expected from Warburg's, the bank of which he was a director, Lord Gladwyn explained he had called the meeting in his personal capacity to help 'an old friend – Jules Moch'. But the questioner persisted, 'my invitation to this meeting was written on Warburg's headed note-paper!' Needless to say, the meeting produced neither an anti-Tunnel organisation, nor the money to finance a campaign.

The studies of the Anglo/French Working Group of Government Officials, which lasted over two years, were at last published on 3rd September, 1963. They were released in Britain as a White Paper[34] which concluded in favour of the twin railway tunnels recommended by the Channel Tunnel Study Group. In summary, the conclusions of the Report were that the bored tunnel project was technically feasible, and that passenger safety arrangements were adequate. The estimated traffic flows could be accommodated, and from the economic point of view the Tunnel would be less costly than a bridge. The economic returns on the Tunnel programme appeared satisfactory, and compared favourably with returns on other transport investments. The Report concluded that the 1960 cost estimates of the Channel Tunnel Study Group's proposal should be increased by 25%, raising the total cost to £160 million, including financial charges. This increase was principally derived from a revision of the reserve for contingencies.

The Official Working Party did not, however, favour the Study Group's novel idea of financing the Tunnel privately and leasing it at a nominal annual rental to the two Governments. It was thought that the method of direct guarantee of loans would be preferable to the granting of special fiscal privileges.

The main criticisms of a bridge project revolved around the inconvenience to shipping and its high capital cost, £298.5 million, nearly double that of a Tunnel, without the increase in traffic to compensate.

There was, however, an incident, which marred the occasion. On the eve of the White Paper's publication, Channel Tunnel shares stood at some 17 shillings, having for no apparent reason fallen by a similar amount during the previous week. The White Paper was

---

[34] Command 2137, *Proposals for a Fixed Channel Link*, HMSO 1963.

not officially released until the afternoon of 3rd September, but when the London Stock Exchange opened that morning, the shares began to soar, and over the next two days rose to over £4.00. This dramatic rise in value prompted a flurry of questions. Had there been a leak? Calls for an enquiry came from the floor of the House of Commons. The stockbrokers concerned, however, put a stop to any further rise by putting a premium on dealings. The excitement died down, no enquiry was forthcoming, but rumours persisted.

Leo d'Erlanger was very distressed by this episode, as his overriding concern throughout was that the private financing should be beyond reproach. At every Annual General Meeting he would deplore any speculative dealing in the shares. After all, the Company had never paid a dividend and its only asset was a hole in the ground.

The White Paper generated considerable interest and excitement, but contributed little to the actual progress of the venture, apart from a fast-growing groundswell of favourable comment in the media.

POINT COUNTERPOINT 1964 – 1970

On 4th February, 1964, Ernest Marples and his French equivalent Marc Jacquet, *Ministre des Travaux Publics et des Transport*, issued a joint statement:

> . . . the construction of a rail Channel Tunnel is technically possible and in economic terms it would represent a sound investment of the two countries' resources. The two Governments have therefore decided to go ahead with this project.

Channel Tunnel Survey, 1964-65. From left to right: Harold Harding, Leo d'Erlanger, René Malcor, Alfred Davidson.

It was one of the most important statements in the latter-day history of the Tunnel. President de Gaulle, with his usual grasp of momentous events, dispatched a telegram to HM Queen Elizabeth II which echoed Napoleon's famous dictum of 1802:

> The French are deeply aware of the historical importance of the decision. I am convinced that Great Britain and France will find strong reasons to reinforce still further their solidarity and friendship.

Her Majesty replied:

> I thank you very warmly for your friendly message regarding
> the Tunnel under the Channel. I am sure that the carrying out
> of this important project will have happy long-term results for
> the peoples of our two countries.

With de Gaulle – who had two years earlier vetoed Britain's first
application to join the EEC – as an enthusiastic supporter, the
project was rapidly attracting a following wind of approval. The
Export Council for Europe foresaw the speeding up of Britain's flow
of exports to the Continent. The National Union of Railwaymen
welcomed the decision. The Automobile Association expressed the
hope that the Tunnel scheme would include large-scale road
improvements in South-East England. Only the Road Haulage
Association was dismayed that there should be no road-link across
the Channel.

However, despite these high-level exchanges of intent and
goodwill, an official approach had yet to be made to the Channel
Tunnel Study Group which had spent well over one million pounds
in preliminary studies. Leo d'Erlanger drew attention to this irksome
delay at a press conference held by the Study Group in March 1964
to launch its new publication *The Channel Tunnel – The Facts*. This
quoted extensively from the White Paper and presented a
conclusive case for the construction and private financing of the
project. It was prominently featured by the international press. Many
newspapers bluntly asked: 'Why the delay?'

The press conference and consequent publicity was
considered in civil service quarters to be provocative and untimely –
nevertheless it provided the necessary spur, for the date of the first
formal meeting between the Study Group and officials of both
Governments was agreed within a matter of days.

Now that the Study Group was enjoying close consultations with
the two Governments and officials, events moved apace. On 1st
July, 1964, it was announced simultaneously in London and Paris
that the Governments had agreed jointly to supervise and finance a
geological survey to be conducted by the Channel Tunnel Study
Group. This decision should not influence later governmental
deliberations, but the cost of the survey – some £1.25 million –
would eventually be charged to 'whatever organisation may be set
up for the Tunnel itself'.

René Malcor and Harold Harding (now President of the
Institution of Civil Engineers), who as chief consulting engineers to
the Study Group had been responsible for the Group's site
investigations and feasibility study in 1959, were to direct the new
survey. This followed the 'classic' route recommended in the Study

Group's Report. The *Commission de Surveillance* appointed by the two Governments would be directed by Jean Mathieu, representing the French Ministry, and Colonel McMullen, Chief Inspecting Officer of Railways for the Ministry of Transport. The work would be co-ordinated, recorded and interpreted by the Study Group's consultants[35].

Thirty-one contractors were invited to tender and in conjunction with the two Governments the final choice was made on a strictly competitive basis and contracts were awarded on 30th July, 1964[36].

The marine core-boring programme, representing nearly three-quarters of the total cost of the survey, included the sinking of some seventy bore-holes into the sea bed, a further ten land bore-holes in the Dover/Folkestone area, and seven in the area of Sangatte. It was decided that the core-boring programme would be undertaken from World War II converted tank-landing craft, with specially constructed drilling platforms built out over the side.

During the autumn and winter of 1964 the survey was hampered by particularly bad weather, further complicated by two of the vessels – *Servitor* and *Uniserve* – proving unsuitable. Despite the Herculean efforts of the *Cauville* and *Grosvenor* (renamed GW

The Gem III *carrying out part of the core-boring programme.*

14), the number of completed boreholes was pitifully small, with little hope of the work being completed within the scheduled twelve months. After an exhaustive worldwide search, the Study Group acquired two oil-drilling platforms: the *Gem III*[37] and the *Neptune I*[38]. Harold Harding wryly commented: 'It was like using a sledgehammer to crack a nut!' Suddenly, the next year, the capricious Channel lay calm as a millpond for much of the summer and the survey was completed on schedule.

While it was in its final stages, the

[35] Sir William Halcrow & Partners; Livesey & Henderson; Rendel, Palmer & Tritton; *Société Générale d'Exploitations Industrielle*; and the *Société d'Études Techniques et Économiques*.

[36] Contracts were awarded for the marine boring programme to Wimpey Central Laboratory of George Wimpey & Co. Ltd. and Forasol of Paris, who combined to form the Wimpey-Forasol Limited Venture. The contract for the position-fixing equipment was awarded to the Decca Navigator Co. Ltd. and the contract for the sonar survey was awarded to Egerton, Germerhausen and Grier, Inc., of the USA.

[37] A four-legged section of the ill-fated Sea Gem which had capsized.

[38] A huge platform supported by three lattice girder legs and capable of drilling to depths of 20,000 feet. At the time it was undergoing commissioning trials prior to drilling surveys in the North Sea for Total Oil Marine Ltd.

Prime Minster, Harold Wilson, received an all-party deputation from
the Channel Tunnel Parliamentary Group, who stressed the need for
an early decision. It was important that the teams who had worked
on the project for the past six years should not be disbanded. The
Group was not satisfied with the present rate of progress and urged
an announcement from the two Governments as to what type of
organisation would be entrusted with the project. The Prime
Minister replied that he foresaw no reason for delay in moving on
to the next stage, and that he was ready to take steps to see that the
question of finance be addressed immediately.

The Parliamentary Group had grown impatient. Nearly two years
had passed since the Joint Governments' declaration in principal of
February 1964 and there was still no statement on financing. It was
conjectured that, whilst the British were opposed to any major
Government expenditure in view of the economic situation, the
French were in favour. President de Gaulle was himself profoundly
against a further influx of American capital participating in French
industry. On 30th September, 1965, at a press conference at Dover
Castle to mark the conclusion of the physical aspects of the
geological survey, Leo d'Erlanger said:

*Channel Tunnel
Survey, 1964-65.
From left to right:
René Malcor,
Professor
Bruckshaw, Harold
Harding inspect
samples of chalk.*

> I am certainly against State enterprise alone, but I have veered
> away from private enterprise alone. I think the idea would be a

combination of the powers that can only be exercised by Governments – allied to the resourcefulness and energy of private enterprise.

Marc Jacquet, French Minister of Transport, addressed the National Assembly ten days later: he, too, could express no more than his hope that a decision would soon be reached:

> A complete report on the survey, prepared jointly by the British and French experts, will recommend the solutions best adapted to meet the technical, legal and financial aspects. This joint report should be submitted at the end of this year, or the beginning of 1966 . . . I think, moreover, that the foreseeable excellence of the project and its relative ease of construction – technically speaking – will inspire the legal and financial experts to find the necessary solutions.

In the late Spring of 1966 the Study Group's consultants completed the analysis of their survey, and submitted to the two Governments a detailed report which concluded that there were 'no major adverse geological conditions which might impede the construction of the Tunnel, whether bored, or immersed.'

At an outturn cost of £2.1 million the survey had confirmed the beliefs and assumptions which had been obtained by the crude methods employed in the early nineteenth century. However, guesswork was no basis for a project of such magnitude.

Barbara Castle, Britain's first woman Minister of Transport, announced on 29th June, 1966, that the official report on the organisational and financial aspects of the Tunnel was nearing completion, but it was 'unlikely that site work on the construction could begin before 1969, so that the Tunnel would not be operational before 1974, at the earliest'. The following day, the Prime Minister declared that 'the results of the geological survey are more encouraging perhaps than anyone could have hoped, and the latest estimates of the economic possibilities seem to be more encouraging than the last time such estimates were made.'

During the following week, on 8th July, the French Prime Minister Georges Pompidou came to London for talks with Harold Wilson. They covered a wide range of mutual interests and common problems. According to informed sources they failed to reach agreement on most of the issues discussed. However, in order to project an impression of unity, the two Prime Ministers gave their blessing to one topic on which they were very much in accord:

> It was agreed that the report of the Commission, after careful examination on both sides, would be discussed between the two Governments with a view to finding a solution for the

construction work on mutually acceptable terms. Subject to
finding such a solution, the two Governments have now taken
the decision that the Tunnel should be built.

This new statement of intent did not arouse any great public
interest; it had all been heard many times before.

Following General de Gaulle's declaration *'Je suis tunneliste'* it
was hardly surprising that the project continued to enjoy support in
France. As with so many of the enigmatic attitudes of the General,
there had been frequent conjecture as to the reasons for his
enthusiasm. The Tunnel could have been the symbolic gesture he
was seeking that signified Britain's commitment to Europe, resolving
any doubts as to her readiness to become a full member of the
European Community. It would also most certainly be the salvation
of the economically-depressed *Pas de Calais* region, and thus would
fulfil his pledge to his brother-in-law, Vendroux, the Mayor of
Calais. Vendroux had often said that he relied on Madame de
Gaulle to keep her husband acquainted with Calais's desperate
need for development – what better than a link with Britain 'since
there is so much English blood in the veins of the *Calaisiens*'?

Whatever his motives, the General would countenance no
interference with his resolve that the Tunnel would be built. In 1964
he underwent an operation and in his absence the French Cabinet
held its regular Thursday meeting. They summarily rejected the
Tunnel on the grounds that no provisions had been made in that
financial year. Immediately following this it was announced that all
meetings of the Cabinet were cancelled until the General returned
to active duty. The story continues that the General, on his return,
rebuked his Ministers for their precipitous action – they should not
have been so deterred by the lack of itemised funds as private
capital was readily available.

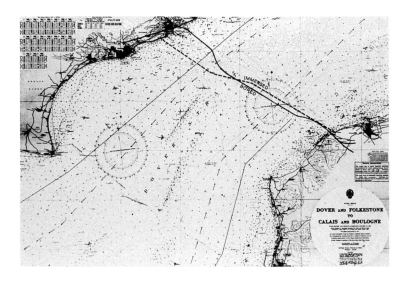

*Chart proposing routes for
bored and immersed tube
tunnels, 1960s.*

The task of 'finding a solution' was now handed over to Barbara Castle and Edgar Pisani, the new French Minister of Transport. After a conference in Paris, they issued the following joint statement on October 28th, 1966:

> . . . the British and French Governments have agreed the lines of a solution for the provision of the Channel Tunnel. The Tunnel would be built with recourse to finance from private sources, but its operation would be entrusted to an Anglo-French Public Authority.
>
> This announcement follows the agreement of 8th July between the two Prime Ministers, M. Pompidou and Mr. Wilson to the effect that the two Governments had decided that the tunnel should be built, subject to finding a solution for the construction work on mutually acceptable terms.

The Channel Tunnel Study Group had proposed, in 1960, that the Tunnel should not only be built by private capital, but operated by a private company; however, it accepted this new proposal as a practical solution. The repayment of the private investment for construction would be made through a form of royalty arrangement on all goods and traffic using the Tunnel. A new estimate by the Study Group now put the total cost of the Tunnel at some £200 million at 1966 prices.

Another four months passed. In February 1967 the two Ministers of Transport dealt what at first appeared as a rebuff to the Study Group. Their announcement said that the private financing of the project would be put out to tender, and all interested financial consortia were invited to 'make themselves known'. This seemed a somewhat churlish return to the Study Group for its ten years of endeavour at a cost far in excess of the million-pound mark. However, taking courage from the fact that its members were more likely to know the problems and difficulties involved than any callow newcomers to the scene, they were the first to 'make themselves known' and were shortly joined by two competing consortia.

Together the three groups comprised an impressive array of 25 international financial houses, including, ironically, several who had in earlier years trivialised the Study Group's endeavours to promote 'that crackpot scheme'. Their first task was to submit, by July 1967, their plans for private financing and management of the construction of the Tunnel. In November it was revealed that the consortia were working closely with French and British officials, elaborating their original submissions, and that the Governments expected to select a group for the construction by the New Year, a forecast which would enable work to start in 1970 and finish in 1975 or 1976.

Why the Governments should have opted for this procedure

has remained somewhat obscure – unless it was to delay matters. Doubts were expressed at the time – later justified – as to the value of going through this competitive financial exercise, as such proposals would be of little value, based as they would be on 1966 prices, for a Tunnel to be built in 1970.

The 5th of April, 1968, witnessed Harold Wilson's long-awaited Ministerial re-shuffle. Barbara Castle was elevated to the relatively quieter pastures of the newly-formed Department of Employment and Productivity and was succeeded at the Ministry of Transport by Richard Marsh (later Lord Marsh), who came from the Ministry of Fuel. His ready wit and his common sense approach no doubt lightened the workload of the office which he had assumed – not the least of which were the closing phases of the mammoth Transport Bill, which had been introduced by his predecessor. This included a section which enabled the Minister to acquire land which was likely to be needed for the provision of construction work on 'a railway linking England with France'.

Three weeks before the close of the final committee stage of the Transport Bill, Richard Marsh introduced an enabling clause entitled the 'Channel Tunnel Planning Council'. The function of the Council was to do the necessary preliminary work before the building of the Tunnel was ratified by Parliament.

But the House was in no mood to accept the introduction of a major Bill without proper warning and erupted in a volatile exchange of views. 'This is the most appalling manner to treat Parliament', declared Peter Walker, 'there will be no time to discuss this measure . . . this is turning Parliament into a Reichstag'. Other members protested equally strongly, with Labour back-benchers retaliating with shouts of: 'Rubbish!' – 'synthetic indignation', and 'you ought to be ashamed of yourselves'. All good knockabout stuff, well in keeping with the traditions of the House.

Richard Marsh's unruffled rejoinder was to express concern for the Opposition members who had given tongue before they had studied and understood the new clause. Emphasising that it in no way committed Britain to the building of the Tunnel, he hoped to be able to select soon, in conjunction with French colleagues, a group which in due course would form the construction company for the Tunnel. His contribution was eminently practical, since it provided Government departments with the necessary statutory powers, and gave their officials purpose and direction.

During this brief ripple of Parliamentary interest in Tunnel matters, the French Government was preoccupied with an upsurge of civil unrest. May 1968 saw the Sorbonne students in a state of revolt. The unrest rapidly spread across the nation: strikes proliferated. Factories, shipyards, radio, TV and postal services were curtailed, rail transport disrupted and traffic in city centres frequently brought to a standstill. When workers began to take over a number of companies, it looked as though France was on the brink of a new revolution. A General Election was held in June, and a financial crisis developed. Three budgets, with tax increases, were passed by the General Assembly: in July, November and December. Confidence in the franc was shaken, and de Gaulle had to assure the Nation and the world that there would be no devaluation.

Understandably, while France was undergoing such a painful process of reassessment, no progress was made on the Channel Tunnel.

Hovercraft preparing to leave Calais.

The first hovercraft cross-Channel service operated by British Rail Hovercraft Limited was scheduled to start in August. Its inventor, Christopher Cockerell (later knighted) told *The Times* that a service, costing a mere £20 million, could do the same job as the Channel Tunnel at one tenth of the cost. This was hardly fair comment. His enthusiasm caused Cockerell to gloss over the fact that, although the Tunnel might well cost ten times as much as the hovercraft, it had in excess of twenty times the capacity irrespective of weather conditions.

The hovercraft commanded a powerful lobby in Westminster, and questions were asked in the Commons. Richard Marsh replied:

> Before any final commitment to construction of the tunnel, there must be a further review of its viability and for this review new cost and revenue studies will be made. Hovercraft, as one of the methods of cross-Channel transport[39] will be among the factors in these studies.

The situation in France had eased by September and President de Gaulle told the retiring British Ambassador, Sir Patrick Reilly:

> The British Isles have not ceased to be distinct from the Continent, particularly from a certain Common Market, but for what is essential, I mean mutual esteem and friendship, cooperation in important fields of progress should I mention the Jaguar (aeroplane), the Concorde, perhaps soon the airbus and one day the Channel Tunnel? . . . I do not indeed think that our two peoples have ever been closer than they are now.

---

[39] This unique form of transport never fulfilled its exciting early potential as a land vehicle with a military application for travelling over difficult terrain. Classified as being neither ship, nor aircraft, with its captains requiring qualifications in both modes of transport, it could never become a major challenge to the sea ferries, let alone the Tunnel, vulnerable as it is to weather conditions.

At the 88th Annual General Meeting of the Channel Tunnel Company, the usually optimistic d'Erlanger observed, 'the mills of God grind slowly, but when the mills of two Governments grind, I suppose they rarely get out of first gear; and so we must possess ourselves in patience'. Prophetic words, for when Richard Marsh reported to the House nothing could have been more disheartening:

> The French Minister of Transport and I have completed our consideration of proposals for the financing of the Channel Tunnel submitted by three competing private groups earlier this year.
>
> We have noted with satisfaction the considerable progress made towards defining the technical and financial terms for a jointly acceptable solution. We are also agreed that certain aspects of all the proposals need further improvement, and that we should not be justified at this stage in narrowing the field for further negotiations to any one group of the three.
>
> We have, therefore, established jointly agreed guidelines for the use of the groups, who are being invited, in the light of this further guidance, to supplement their earlier proposals, or should any of them prefer to do so, to combine for the presentation of fresh proposals.
>
> We are confident that this final stage of talks will reach a successful conclusion leading rapidly to the choice of a private group to carry out further studies required for final assessment of the project and, given favourable results, to arrange for the Tunnel to be constructed.

This brief joint statement indicated a prevailing indecision. Over twenty months had passed since Barbara Castle first invited financial tenders – only for the three financial groups to be told that they would have to start all over again! The sceptics, who had declared that little purpose would be served by engaging in a financial knock-out competition, claimed no satisfaction for being proved so right. As had been envisaged, their respective financial formulae in 1967 money terms bore no relation to the project.

The two Ministers indicated that a merger between at least two of the banking consortia would not be inappropriate, and had the grace to appreciate that, in order to avoid another 'hit-and-miss' affair, they would have to lay down clear guidelines. At a meeting with the Channel Tunnel Parliamentary Group on 18th November, a slightly more encouraging view was given:

> The lack of choice at this stage [Richard Marsh said] is not a reflection on the abilities of the three groups. We are all faced with a problem for which there is no comparison in past experience. Thoroughness now will shorten negotiations later.

He went on to say that a complete engineering design as a basis for tenders was necessary and, while this was being carried out, traffic

Frank Davidson,
President of Technical
Studies Inc., New York,
and a full member of the
Channel Tunnel Study
Group, describes the
project to an interested
spectator.

and revenue studies could be completely revised to bring them up to date and take account of latest developments in other forms of cross-Channel transport, including container ships and hovercraft.

> It is unlikely [he concluded] that these will have seriously reduced the considerable economic lead that the Tunnel project enjoyed in earlier studies. But before committing ourselves to a project costing over £200 million, we must be absolutely sure.

Although it was rumoured that Couve de Murville, the new leader of the Government, was not enthusiastic, the French Minister of Transport said that work could start on the Tunnel late in 1971, and this seemed to be in step with the progress Marsh was making in Britain.

General de Gaulle frequently called for a referendum on national controversial issues, regarding any result contrary to his views as a vote of no confidence. On 27th April, 1969, the referendum on radical reforms in Central and Regional Government resulted in an outright rejection by the French electorate. It was also the rejection of their President – he resigned the same day.

'They've called the old man's bluff', was the comment on both sides of the Channel. The champions of Britain's entry into the EEC felt satisfaction that their chief adversary had now gone. Channel Tunnel supporters, however, had lost one of their staunchest allies. After a short interval, Georges Pompidou – a declared *'tunneliste'* – emerged as the new President.

In Westminster Harold Wilson was said to be urging his Minister of Transport to speed up the Tunnel project as part of his drive to get Britain into the EEC. The foremost task appeared to be to get at least two of the financial consortia to merge and form – in the words of Leo d'Erlanger – 'as broad and widespread a base as one could wish for'. Richard Marsh's response was to announce in July that, having taken into consideration the opinions of regional and local authorities, the shuttle terminal would be sited at Cheriton, just outside Folkestone, with a Passenger Station at nearby Saltwood and a Freight Yard near Westenhanger – the whole scheme to be known as the 'Cheriton Package'.

Earlier in May an important congress had been held in Lille. One of the speakers was Roger Macé, who led the French inter-ministerial group for the study of the Tunnel. A report of the congress was given in a letter published in *The Daily Telegraph*, from Conservative MP Geoffrey Wilson, the honorary secretary of the All-Party Channel Tunnel Parliamentary Group:

> What impressed me most about the conference was that the representatives of the French, Belgian, Dutch and German

Chambers of Commerce described themselves as belonging to '*la Communauté des Régions de l'Europe du Nord-Ouest*', which they clearly regarded as being a single industrial complex with common interests to which they were anxious to add the industrial areas of Britain.

They favoured a railway tunnel under the Channel for this purpose and did not seem to regard hovercraft, aircraft, or roll-on, roll-off ferry ships as a sufficient alternative to meet all industrial needs, and the Germans, in particular, were interested in the suggestion that trains might run between central England and central Europe.

Far from expressing any expectation that a Channel Tunnel would divert overseas imports from British and French ports, the only doubts expressed were those of the Dutch who feared that a tunnel might encourage imports at the Port of London, rather than Rotterdam, but the Dutch representatives were prepared to take that risk in order to secure an enlargement and strengthening of the whole industrial complex.

Harold Wilson, one of the Prime Ministers responsible for the second attempt to build a Channel Tunnel.

The British delegation, particularly the members of Parliament attending, were impressed to find that on the Continent people confidently expected the Tunnel to be built within five years, despite the economic situation which was still causing such concern in both Britain and France.

In the Commons, Harold Wilson was pressed to say definitely whether he favoured the 'immensely costly Channel Tunnel project'. He made the point that 'other services, such as the hovercraft, could not be a substitute for the Tunnel with all its likely economic benefits. However, a great many problems still had to be ironed out . . . we must get it right before a final decision is taken. Recent press speculation has surprised me, saying I was rushing ahead. It is very much wide of the mark.'

It was indeed 'very much wide of the mark' – 'rushing ahead' was a totally unrecognisable movement to the patient tunnellers.

On 25th July, the tenth anniversary of the first Channel crossing by hovercraft, Sir Christopher Cockerell spoke about the future of cross-Channel traffic:

> The Tunnel would make a loss unless subsidised. From the passengers' and taxpayers' point of view there should be competition . . . over the next 50 or 60 years. By that time, Channel traffic will have increased sufficiently to make the building of a bridge, rather than a Tunnel, an economic and viable proposition. A bridge is the ultimate link and, unlike the Tunnel, can be adapted to accommodate new forms of transport. It could be arranged, for instance, to carry a hovertrain offering a journey time of one hour between the centre of London and the centre of Paris.

It apparently did not occur to Sir Christopher that the Tunnel could also accommodate hovertrains.

In the same month Neil Carmichael, the new Parliamentary Secretary to the Ministry of Transport, announced yet another deadline for the selection of the financing group: 'before the end of the year'. Negotiations in London and Paris would begin 'soon', and the interested groups had been 'encouraged to move closer together with a view to sharing the load'.

The *Financial Times* (28th July) remained sceptical. The build-up for the project had become as slow-moving and as interminable as a 'soporific TV soap opera'; 'anything dramatic, or decisive, always seems as far off as ever'. As for financial groups 'getting together in a single monster consortium', a member of one of the groups had remarked: 'Having organised competition, the two Governments now appear to be embarrassed at having to choose, and would like to be helped out of their predicament'. Before a merger could take place compensation to the Channel Tunnel Study Group would have to be resolved, and 'who should be in the saddle in a consortium of twenty-five bankers and four other companies from four different countries?'

Just before the 1969 Summer Recess, the House spent a full hour's debate on the Channel Tunnel – the most extensive for several years.

William Deedes (later knighted), Member for Ashford, admitted to his local interest when he voiced his concern: 'I am in blank ignorance of the current factors for and against a Channel Tunnel', he said. 'We have had no reliable data from the Government since 1963 . . .' In particular he referred to his understanding that a final feasibility study lasting two years was likely to start in October. 'I am hostile to the project', he admitted, 'so that in my opinion the longer it takes, the better. But two years requires a word of explanation and, since we have been conducting inquiries into the project for some time, it postulates a puzzling lack of urgency.'

Deedes concluded: 'At one time, there were three consortia entering bids against each other. Then they were encouraged to become one. Where has all this got to?'

Albert Costain, the Member for Folkestone, contributed to the debate and gave his own impressions of the Lille Congress:

> The French were extremely enthusiastic about the Tunnel idea. They explained how North France was at the delta of the Rhine, and how the axis of communication would be completely altered by the Tunnel . . .
>
> We were given some interesting figures which, for some reason, have still not been published in Britain. For example, we were told that Channel-crossing foot passengers were increasing at the rate of 1.5% a year, that Paris-London air traffic which, in 1967, totalled 1,235,000 passengers, was increasing by 9% a year, and that the tonnage of merchandise being carried

was increasing by 4% a year. The most startling figure of all was
that motor-car traffic was increasing by 14% a year.

Costain, despite the growing concern in Kent and in particular
amongst his own constituents, had always been a staunch supporter
of the Tunnel. He was brother to Sir Richard Costain, Chairman of
Costains, one of the UK's largest construction companies, who was
actively promoting the immersed tube tunnel concept. Although
Albert had no association with the company, whenever he made
one of his many favourable comments on the project he was
invariably greeted by Labour back-benchers with chaff such as 'Jobs
for the boys!' and similar ribaldry, but his good humour remained
unruffled.

David Crouch, Member for Canterbury, who held entrenched
views against the Tunnel and believed France would be the
principal beneficiary, said: 'No wonder the French are so keen on
it!' But even Crouch was eventually to undergo a change of heart.

Geoffrey Wilson, Member for Truro, observed that a political
and psychological attitude lay at the root of much of the
antagonism:

> Much of the opposition to the Channel tunnel is due to the
> conflict of two political trends which are spreading around the
> world like an influenza epidemic. On the surface, in the Iron
> Curtain countries, in the newly-emerging countries, and in the
> old democracies, the trend is towards local independence in
> smaller and smaller units and the demand for people to manage
> their own affairs . . . Much of the opposition comes from the
> belief that 'wogs begin at Calais' . . . some of my constituents
> think that 'wogs' begin at Plymouth. Below that feeling there is
> a political trend towards wider economic unity in larger areas,
> and hence we have moved towards the common market, the
> Communists' trade pacts and talk of an Atlantic Community, and
> so on.

Neil Carmichael answered for the Government. He again outlined
the project briefly, gave the 'state of play' and denied that the
Ministry of Transport was withholding information from Parliament
and the public: 'In fact', he said, 'the British Government have kept
the House better informed than the French have the National
Assembly.'

Michael Heseltine, Member for Tavistock and Opposition
Transport spokesman, ended the debate on a note of urgency:

> One point about which the Parliamentary Secretary [Neil
> Carmichael] was not as forthcoming as I would have wished
> him to be, concerns the question of delay . . . The Ministry is
> not keeping to the target which it has published . . . Today, the
> Parliamentary Secretary said that he hoped that agreement [with

a financial group] could be reached before the end of the year. He merely 'hoped' that agreement could be reached. It is possible that agreement to proceed with the detailed study will not be reached until the beginning of 1970. That puts back the schedule, with a two-year investigation period followed by a five-year construction period. Therefore, we must talk about a Channel Tunnel coming into operation in 1977 or 1978, with the immediate effects which it will have on any cost projections, but it would be helpful if the Ministry could inject into this early negotiating stage a degree of urgency.

This extensive debate provided much information on the feelings of MPs and their constituents towards the Tunnel, but Richard Marsh unfortunately had no opportunity to act on the demand for urgency. As the result of another Government reshuffle in early October 1969, he was replaced by Fred Mulley. In Paris, too, the Transport Ministry had changed hands: in June, Raymond Mondon, who was also Mayor of Metz (Moselle) had taken over from Jean Chamant. A cheerful note of prophecy was sounded by Pierre Decoster, President of the Lille Chamber of Commerce. 'The Tunnel will cost less to build than what France pays out in butter subsidies in one year', he said. 'Almost every tunnel ever built cost less than at first estimated, and paid for itself earlier than people expected.'

Population figures in millions of the area likely to take advantage of the full benefits of the Channel Tunnel by 1978.

The optimism and enthusiasm of the people of north-west France for the Tunnel had to be seen against the economic background of post-war France: the Calais-Lille corner of the country was declared a development area – the coal-mining and the traditional textile industries were being cut back and new employment had to be created for a growing pool of labour. Besides, Lille lies closer to Dover than to Paris.

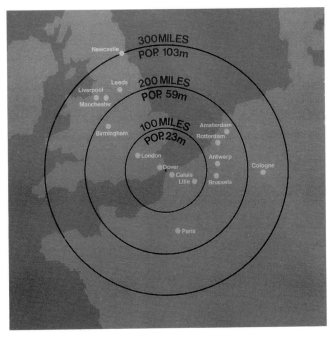

Unexpected support for the Tunnel came from America at the end of 1969 from Major-General Leif Sverdrup who confounded the bridge and the bridge-tunnel-bridge promoters. They had regarded him as a firm ally because he had designed the bridge-tunnel-bridge link across Chesapeake Bay; but, at a meeting at the Institution of Civil Engineers in London, he surprisingly favoured a bored tunnel He was totally opposed to the Chesapeake solution, and he declared 'the building of a bridge across the Channel would be a very hazardous operation indeed'.

As the year, with all its peaks of

enthusiasm and valleys of despond, drew to a close, Fred Mulley and Raymond Mondon met in Paris for what was described as 'an informal chat about the progress made on the financing of the scheme'. At the end of the meeting there was the usual 'soothing joint statement': the 'go-ahead' for the Tunnel, was 'likely to be made in January 1970'.

In true Hogmanay spirit, the press was optimistic: 'Digging will start in the New Year', said the *Sunday Express* and 'the great machines that will bore their way between England and France are expected to be given trial runs.' The *Evening News* compared the coming 'engineering marvel' to the 'daily experience of Londoners':

> Off to the Continent . . . by underground. London to Paris in four hours in one train. It could be a fascinating adventure for millions. But it will be old hat for regulars who travel on the Tube between Morden and East Finchley. They are quite used to subterranean journeying. They have been rattling along the world's longest continuous tunnel for years – 17 miles of it. The Chunnel will be 21 miles long[40].

By January 1970 the three banking groups had agreed in principle on a merger, but the two Governments' insistence on detailed safeguards – to ensure that the concession would never become a licence to print money – was believed to be holding up final approval. *The Times* was expecting the merger to be made public 'before the Spring'.

Five years had been wasted in a tedious repetitive spectacle – the consequence of indecision, indifference and procrastination – coupled with a total lack of imagination.

While there was widespread impatience with the lack of progress, and with the Tunnel's popularity waning fast, it was something of a paradox to witness the early beginnings of organised opposition in Kent. Hitherto, criticism had been limited to the preservation of the status quo by vested interests. An organised anti-Tunnel lobby would appear to indicate that the project was beginning to make ground – but there was precious little evidence to support this view.

THE NEW INTERNATIONAL GROUP

The Channel Tunnel Company held its 89th Annual General Meeting on 15th July, 1970, some three weeks after the return of the Conservatives to power, under the Premiership of Edward Heath (later knighted). The small group of loyal shareholders heard their Chairman conclude his statement with his usual words of comfort.

---

[40] Possibly a typographical error on the part of the *Evening News*: the Tunnel will be 31 miles long.

Edward Heath, one of the Prime Ministers responsible for the second attempt to build the Channel Tunnel.

Suddenly their patience was rewarded. Leo d'Erlanger had been handed a message which he read aloud:

It is announced that members drawn from the three groups invited by the Ministry of Transport to submit proposals for the finance and construction of a Channel Tunnel have joined together to form a new group, which has today submitted joint proposals to the British and French Governments. These proposals will cover a basis for the financing and conduct of final studies leading after some two years to the final decision whether to build the Tunnel, and the terms on which, after the studies, the group would be ready to finance the works and arrange for their construction, which would take about five years. If the proposals were accepted this year the Tunnel might be open in the late 1970s.

The new group includes: in the United Kingdom – The Channel Tunnel Company (in which British Railways are major shareholders), Morgan Grenfell, Robert Fleming, Hill Samuel, Kleinwort Benson and S G Warburg: in France a number of banks under the leadership of *Compagnie Financière de Suez, Compagnie du Nord, Banque Louis-Dreyfus* and *Banque de Paris et des Pays-Bas* together with French Railways. The Group will also include some American investment banks, details of which will be announced at a future date.

Conditioned as they were to setbacks, the shareholders received this news with equanimity.

Later that day the statement was officially confirmed in the House of Commons by the new Minister of Transport, John Peyton. However, beyond the names of the participants, the two Governments revealed nothing of the new financial group or its proposals. In fact, the new group could not officially exist until the Governments responded to its submission.

An acceptance of the proposals would eventually lead to an agreement; but each Government differed on how to achieve the complicated partnership between private and public enterprise which the project would require. 'Persuading the French to agree to anything is not an inconsiderable difficulty', said ex-Minister of Transport, Richard Marsh, in a radio interview on the day of the announcement:

There are some very real problems here. First . . . a method of raising money. Secondly, British Railways and French Railways formed into a company to operate the Tunnel. Thirdly . . . set up an Anglo-French company to own the Tunnel . . . the other thing, of course, is getting the last up-to-the-minute survey of how much it's really going to cost, how much it's going to produce. There's no sense in having a Tunnel for the sheer fun of the thing – you want it to pay . . . there have been two previous studies . . . the second one showed a big increase in the profitability on the first . . . if the third shows that it's profitable, then we'll pull the stops out and go ahead. But the main thing is that it is pretty daft that it takes people so long to cross this tiny strip of water. I think it is absurd. It's extraordinary that so many people have been prepared to put up with such a primitive method of travelling 20 miles for so long!

Leo d'Erlanger was also interviewed but his experience of the French was quite the reverse.

I can assure you that, where we have opposite numbers, French and English, at every level it's quite extraordinary the measure of harmonious collaboration that we have had. I hope now we might see something beginning to happen. You never know, I might see the Tunnel built before I die.

The announcement of 15th July, low-key as it was, would decide the destiny of the Tunnel.

In November 1970 the British Government took a radical decision to combine three ministries – Transport, Housing and Local Government – into one vast, new administrative body: the Department of the Environment. The head of this new Department, Peter Walker, assumed ultimate responsibility for the Tunnel project, but administratively it remained in the care of the Minister for Transport Industries, John Peyton, and it was he who fielded the frequent questions in Parliament with assurances that a statement, a decision, a new development concerning the project would be forthcoming 'before long', 'in the near future', or 'next year'.

As usual, the French seemed more sanguine. In December 1970, the Cabinet gave its approval for certain studies to be carried out, and Pompidou told his Ministers that the Tunnel would 'transform Britain's relations with the rest of Europe – and above all with France'.

At the end of January 1971, John Peyton had something more concrete to report, having been to Paris for weekend talks with Jean Chamant, Minister of Transport. They had gone over the proposals which had been submitted to the Government by the international banking group six months earlier. 'Both Governments', Peyton declared, 'will now discuss the proposals and possible modifications of them with the group before reaching conclusions. We shall also be discussing with the group how an early start can be made on the further studies which must precede the final decision on the project itself.'

The French press, however, had another interpretation of what was going on behind the scenes. Some papers reported, after Peyton's visit, that the British Government was placing obstacles in the way of the project by demanding new studies.

There was definitely a new awareness in Tunnel-related matters. In March 1971 Sir Eugene Melville was appointed to a new post in the Department of the Environment, that of Special Adviser on Channel Tunnel Studies. His particular responsibility was the final studies on which the decision on construction would be based. John Peyton told the Commons that the final studies, estimated to cost up to £12 million, were planned to be put into effect in the Summer.

Lancaster House, the official Government reception centre, witnessed a major step forward on 22nd March, 1971. Jean Chamant came to London for Tunnel talks with John Peyton. Their meeting ended in amicable accord and set the seal on their decisions. They were later joined by the French and British representatives of the international finance group. The two Ministers and the group agreed on the Heads of Terms for the final studies and the financial basis on which the Tunnel would subsequently be built and operated.

The Tunnel would be financed by an equity-bond mix – partly from private equity risk-capital (10% to 30%) and partly by Government-guaranteed bonds. Private sector Companies set up by the Group in Britain and France, would be responsible for the financing and construction; they would hand over the completed Tunnel to an Anglo/French Public Operating Authority and be remunerated from its revenues. The Companies would receive a share of profits and a degree of control over the commercial policy of the Tunnel.

Georges Pompidou.

The final studies were to be in two stages. The first would include traffic and economic surveys up-dating the cost and re-evaluating previous favourable assessments of the project, and would take some twelve months to complete. The acceptance by the two Governments of the first stage report would give the go-ahead to the second stage – a continuing refinement of the economic and financial studies, together with definitive technical and engineering studies.

The decision to construct the Tunnel would be forthcoming late in 1973, which indicated a possible completion date early in 1978. The two Ministers announced that they had chosen the group[41] for the further pursuit of the project and that the first stage was to begin at once. At last a solution had been found to a number of sensitive areas of dispute between the two Governments and the private interest. Both sides had every reason to be satisfied with the outcome.

---

[41] See Appendix III.

The new consortium brought about some changes in the *dramatis personae* of the Channel Tunnel Study Group. Merchant banker William Merton, a Director of Robert Fleming, who had successfully led negotiations on the British side, and Leo d'Erlanger still remained, as did Viscount Harcourt, Chairman of Morgan Grenfell – a member of the Supervisory Committee of the Channel Tunnel Study Group since its inception in 1957. Originally representing the Suez Canal Company's interests, Lord Harcourt had become co-Chairman of the Group on the death of Sir Ivone Kirkpatrick in 1964.

Lord Harcourt, Co-Chairman of the Channel Tunnel Study Group from 1957.

Technical Studies Incorporated of New York did not feature in the new consortium. Its members considered that the part they had played in the pioneering role of the Study Group had been accomplished. This meant the withdrawal from the scene of Alfred Davidson, one of the directors. The irrepressible ubiquitous 'Al', as he was known by all, had fulfilled the role of effective liaison, maintaining unified action between members of the Group and Government officials on both sides of the Channel. Consumed as he was with a dogged determination to get 'the show on the road', his single-mindedness of purpose aroused grudging admiration as – undaunted by adversity – he goaded and cajoled those around him, brooking no deviation. For over a decade he had commuted almost weekly between London and Paris. His colourful personality would be missed.

However, the American presence was maintained by three banks – Morgan Stanley & Company, The First Boston Corporation, and White, Weld & Company Limited. These joined what was now officially referred to as 'The British Sub-Group'.

British Rail was now a partner in its own right, whilst still retaining its 26% interest as the major shareholder in the Channel Tunnel Company.

A newcomer to the scene was a company of international repute: the Rio Tinto Zinc Corporation, the British-based parent organisation of the world-wide complex of mining and industrial companies, known as the RTZ Group, under the chairmanship of Sir Val Duncan. In the early 1960s, in common with other national resource developers, RTZ acquired the capacity not only to undertake projects on a vast scale, but the ability to organise and assemble men and machines to build the necessary infrastructure of ports, towns, railways and roads to feed and support such projects. Having forged a high calibre team with a broad spectrum of talents and experience in engineering and construction, RTZ had formed Rio Tinto Zinc Development Enterprises Limited (RTZ-DE) which was to be responsible for the planning, financing and construction of major projects in Europe and those parts of the world where RTZ lacked a sophisticated organisation.

The British Sub-Group (as the Channel Tunnel Study Group before it) had lacked a management and executive structure, and RTZ-DE fulfilled this shortcoming. RTZ-DE was duly appointed Project Manager and entrusted with the responsibility of carrying out the new studies. It, in turn, appointed Cooper Brothers Ltd.[42] to undertake new economic, traffic and revenue studies, and Sir William Halcrow & Partners and Mott Hay & Anderson were appointed Consulting Engineers to carry out engineering and design studies. Building Design Partnership was retained for the terminal design.

The French Sub-Group, under the capable leadership of General Philippe Maurin – who had been appointed Chief of the Air Staff in 1967 – made similar appointments to undertake the studies in France. *Société d'Ingénieurie du Tunnel Sous la Mer* (SITUMER), a consortium of engineering consultants, though not appointed as project managers, was to be responsible for the engineering studies, and *Société d'Études Techniques et Économiques* (SETEC) was to undertake the economic, traffic and revenue studies.

At the Annual General Meeting of the Channel Tunnel Company on 15th June, 1971, Leo d'Erlanger informed shareholders that, immediately following the AGM, an Extraordinary General Meeting would be held to consider two important resolutions: one was to increase the share capital of the Company; the other, to change the name of the Company to 'Channel Tunnel Investments Limited'.

As to the change of the Company's name, d'Erlanger said:

> It has for some time been the intention that the Sub-Groups should operate through corporate bodies in their respective countries. The French Sub-Group is now represented by *Société Française du Tunnel sous la Manche*; and the British Sub-Group is in course of incorporating a company . . .
>
> In view of the role which it is hoped the British Company will ultimately play, we have agreed that it should be called The British Channel Tunnel Company Limited. In order to avoid any possible confusion a resolution will be proposed at the Extraordinary General meeting to change the name of your Company to Channel Tunnel Investments Limited[43].

Both resolutions were passed without dissent, and the British and French Sub-Groups emerged with recognisable identities.

PUBLIC OPINION

In French eyes, the British still did not understand the importance of

---

42 Now Coopers & Lybrand.
43 Under which title it is still quoted on the London Stock Exchange.

the Tunnel, and viewed against the background who could blame them? Roger Macé came to the Royal Society of Arts in Edinburgh to give the Scots a pep talk on the subject. The Tunnel would be beneficial for Scotland, he explained, directly and indirectly, by increasing trade between France and Britain, by improving freight services for the export of Scottish goods – 'including the most famous Scottish export of all.' He went on to say that the Tunnel 'represents something far more important between our two countries than mere economic interests. The Channel is not just a mass of water but a state of mind. It's only a pond, after all, a wee pond as you might say in Scotland!' He concluded by recalling that his fellow-countryman, Ferdinand de Lesseps, made a similar trip to Scotland in 1857 to tell the Society about his ambitious plans for a waterway between the Mediterranean and the Red Sea. Twelve years later, the Suez Canal was opened. History, Roger Macé hoped, would repeat itself.

General Philippe Maurin, CBE, President to the Société Francaise du Tunnel Sous la Manche *and Deputy Chairman of the British Channel Company Ltd.*

The *Pas-de-Calais* – one of France's oldest industrial centres – had become virtually a disaster area as a result of World War II and its inhabitants were suffering from severe industrial and economic blight. Recognising that they were closer to London than they were to Paris, they looked across the Channel for the answer to their problems.

In the belief that sooner or later Britain would join the EEC – but with an even firmer conviction that the Channel Tunnel would be built – they had implemented plans for industrial redevelopment to attract British industry to French soil. During the past few years, some fifteen British companies had responded to the attractive inducements the region offered: selected sites, favourable land prices, building grants, plus local and State tax reliefs for a number of years.

Oliver Le Sourd, the Secretary-Director of *Comité d'Études d'Action pour le Développement Économique de Calais et sa Région* (CEADEC) – the organisation sponsored by the Calais Chamber of Commerce and the Municipality to promote the redevelopment scheme – believed that the development of Calais provided the solution to one of the problems that troubled south-east Kent.

> I can understand [he said] why the people of Kent do not wish to see the 'Garden of England' marred by industry being attracted to the district when the Tunnel becomes a reality. If industry wants to be close to the mouth of Tunnel, then what does it matter which end of the Tunnel they choose? Let them build here – do not spoil your beautiful Kent countryside – the south-east coast can be the dormitory area for British industry in Calais – they can work here and live in Kent. It will be quicker to commute via the Tunnel to Calais than it is now to commute from Kent to London.

No doubt he was sincere, but equally he had an eye to the main chance.

A large area of land in the Calais district had been provisionally earmarked for the Tunnel Terminal complex, bounded by the villages of Coquelles, Fréthun and Peuplinques. Families would have to be re-housed and farmers cultivating some of the richest land in France would have to farm elsewhere. Yet despite the personal hardships involved, the people of the region welcomed the Tunnel, which was expected to provide construction work for some three thousand, and some two thousand long-term jobs thereafter. The enthusiasm of the *Calaisiens* was in direct contrast to the apprehensions felt by some of the inhabitants of Kent. A high proportion were retired, and feared that the Tunnel would end their peace and quiet, and play havoc with the rural environment.

The random voices of opposition in Kent were united on 27th August, 1971, at a meeting in Dover, chaired by Ron Spruhan, Assistant General Secretary of the National Union of Seamen. It was attended by other trades unions, ratepayers' associations, landowners and anyone who feared loss of jobs, amenities or destruction of the environment. The meeting resulted in the formation of the Channel Tunnel Opposition Association.

The National Union of Seamen's participation was understandable, as was Roland Wickenden's – Chairman of Townsend Car Ferries, the thriving cross-Channel sea-ferry company – with his very obvious vested interest in preserving Britain's traditional sea links with the Continent. Wickenden, although vocal in his opposition, did not emerge as a patron of the newly-formed Association, but he was known to be an active supporter behind the scenes.

The Tunnel would provide short-term employment opportunities during the five to six years' construction period and in the long-term the administration and operation of the Cheriton Terminal would make a positive contribution. The very existence of a Tunnel, furthermore, would create opportunities for the area which, if grasped, would provide a new era of prosperity for local industry and coastal resorts. The means of controlling future development, consistent with county requirements, already existed, and with careful planning Kent could prevent runaway expansion scarring the countryside.

A HIGH-SPEED LINK

At the end of October 1971 Parliament voted in favour of joining the EEC: on 1st January, 1973, Britain would at last become a fully-fledged member of the Community. Leo d'Erlanger had always maintained that the Tunnel would be economically essential if Britain joined; but doubly so if it did not, as the reduced delivery times for Britain's exports would enable her to compete more favourably with the rest of Europe. He also repeatedly maintained that, once constructed, the Tunnel with its low maintenance costs would be a permanent buffer against inflation, unlike ships, hovercraft and aircraft which eventually wear out and have to be replaced at future inflated prices.

The idea of a high-speed network of railways throughout western and central Europe, linked to Britain, was slowly taking shape. It is no exaggeration to say that the prospect of the Tunnel had greatly influenced Continental railway planning, in West Germany even more than in France, and it was the Federal German Transport Minister, Georg Leber, who first suggested a unified Common Market railway system which would include Britain after her accession to the Treaty of Rome. Technical problems such as the standardisation of rolling-stock, loading gauge and container dimensions would, of course, have to be solved; but these problems already existed within the original 'Six', whose railway systems had developed independently.

While the project management and independent consultants were at work on their figures and forecasts, British Rail was preparing for the 'Channel Tunnel Age' under its new Chairman, Richard Marsh. Having left his safe seat in Parliament, he accepted the formidable job of trying to steer the railways out of their financial troubles. He may well have sensed that this form of public transport was heading for an international revival with the extension of its passenger and freight horizons into the heart of Europe.

In the late 1960s British Rail had begun its forward planning on the assumption that speeds of 150 m.p.h. would soon be the norm and speeds of up to 250 m.p.h. operationally viable in the 1980's. French and Belgian Railways, too, were of like mind; for example, the Brussels-to-Lille *Europolitan* (later named TGV – *trains 'à grande vitesse'*) line – envisaged as part of the London-to-Brussels high-speed line via the Channel Tunnel – was designed for speeds of at least 180 m.p.h. British Rail's High Speed Train (HST), capable of speeds above 125 m.p.h. was due to enter service in 1976, but Britain's ultimate answer to how passengers would travel tomorrow was the Advanced Passenger Train (APT-E), with a speed of over 160 m.p.h. and scheduled to come into service in the 1980s[44].

In 1969 French Railways also decided on a thorough structural re-organisation. Much of the planning and design work for a faster future owed its impetus to Roger Hutter, Joint Director-General of SNCF. It was during this period of planning and experiment that the potential of tracked hovercraft was investigated by Professor Colin Buchanan; and later, in the Summer of 1971, Britain's prototype hovertrain began its first runs on a test track near Huntingdon. It

The proposed high-speed link between Folkestone and London.

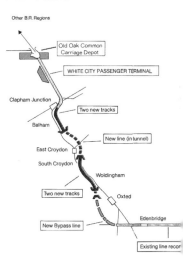

---

[44] See Appendix II.

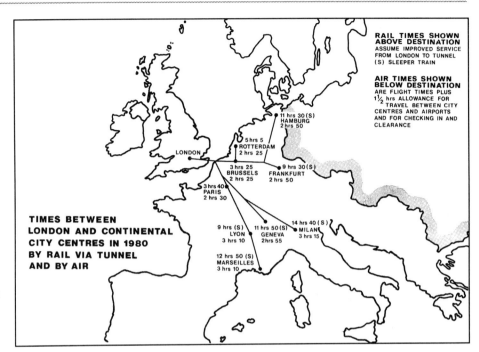

RAIL TIMES SHOWN
ABOVE DESTINATION
ASSUME IMPROVED SERVICE
FROM LONDON TO TUNNEL
(S) SLEEPER TRAIN

AIR TIMES SHOWN
BELOW DESTINATION
ARE FLIGHT TIMES PLUS
1½ hrs ALLOWANCE FOR
TRAVEL BETWEEN CITY
CENTRES AND AIRPORTS
AND FOR CHECKING IN AND
CLEARANCE

11 hrs 30 (S)
HAMBURG
2 hrs 50

5 hrs 5
ROTTERDAM
2 hrs 25

LONDON

3 hrs 25   9 hrs 30 (S)
BRUSSELS   FRANKFURT
2 hrs 25   2 hrs 50

3 hrs 40
PARIS
2 hrs 30

TIMES BETWEEN
LONDON AND CONTINENTAL
CITY CENTRES IN 1980
BY RAIL VIA TUNNEL
AND BY AIR

9 hrs (S)   11 hrs 50 (S)   14 hrs 40 (S)
LYON   GENEVA   MILAN
3 hrs 10   2hrs 55   3 hrs 15

12 hrs 50 (S)
MARSEILLES
3 hrs 10

*Forecast of
journey times in
1980 by rail via
Tunnel and by
air.*

was Professor Eric Laithwaite, however, who contributed the ideal power system for the hovertrain with his development of the linear induction motor. There was also the technique of 'peristaltic propulsion'. Both inventor and engineer alike visualised the future of the hovertrain in relation to its use through the Channel Tunnel; but, however powered, the development of the hovertrain[45] would take many years, and would possibly further delay the decision to build the Tunnel.

Meanwhile, the Department of the Environment's Transport and Road Research Laboratory in Crowthorne, Berkshire, was carrying out studies to establish optimal dimensions for the design of the Tunnel shuttle trains, which would operate from the terminal at Cheriton. The shuttle train wagons would be much higher and wider than existing rolling-stock, but since they would be confined to the shuttle circuit between Tunnel terminals there was no need to conform either to British or Continental loading gauges. The three-day experiment at Crowthorne was carried out with 250 cars, coaches, towed caravans and boat trailers. To preserve effective realism, volunteer drivers were recruited from the general public. Under carefully controlled conditions, they drove on and off the mock-up shuttle trains to establish the best and quickest method to load and unload a train nearly three-quarters of a mile in length. The studies concluded that driving on and driving off could each be accomplished in under ten minutes.

g route reconstructed
g route widened
ute in tunnel
urface route
g railway

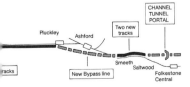

CHANNEL
TUNNEL
PORTAL

Pluckley   Ashford

Two new
tracks

Smeeth
Saltwood
Folkestone
Central

New Bypass line

racks

[45] See Appendix II.

The British and French Channel Tunnel Companies' joint submission for the first stage of the new studies to the two Governments on 9th June, 1972, marked the beginning of an acceleration of events and of heightened public interest.

The completion of the interim results of the studies which were being carried out by the independent consultants, Cooper Brothers and *SETEC-Économie*, could not be regarded as being more than an indication of the outcome of the final study. They did reveal, however, a significant variance in the diverted traffic estimates for 1980 and subsequent years, which in turn affected the overall financial predictions.

SETEC-Économie's traffic estimates were pitched at a fairly optimistic level, compared to the more cautious conclusions of Coopers and Lybrand. The French had based their conclusions on the probability that, when the Tunnel was completed, its attraction would be such that there would be insufficient traffic to support the continuation of the sea ferry services on the short cross-Channel routes. The British on the other hand had assumed that the passenger, traffic and freight forecasts for 1980 and the years beyond were of such proportions that there would be a continuing demand for the ferry services on the short routes – taking into account traditional user-preference, point of origin and destination, and other considerations. The difference between the conclusions of the British and French consultants was subsequently reconciled by a balance being struck between their respective assumptions. The two Tunnel companies concluded in their submission to the Governments that even on the most conservative assumption the financial profitability of the Tunnel was satisfactory, being at least equal to that of similar projects, and significantly ahead of prevailing interest rates.

The interim report demonstrated sufficient promise to justify the completion of the studies, and the Department of the Environment announced its decision on 16th August, 1972, and formalised it by the signing of an agreement between the Governments and Tunnel companies on 20th October, 1972. The agreement provided for the completion, by July 1973, of the economic and technical studies, and would establish a basis for the financing, construction and operation of the Tunnel and a framework for further negotiations. There would be a programme of progress in three phases: Phase One covered the studies to be completed by July 1973, providing all the information to enable Parliament to decide whether or not the project should proceed. Phase Two would embrace preliminary works from mid-1973 to early 1975 and be covered by further parallel agreements. The estimated cost of some £20 million would include the sinking of shafts and the boring of some two kilometres of the pilot tunnel

Leo d'Erlanger.

from each side of the Channel. There would be a second decision break-point in 1975. At this juncture, or earlier, the Governments or the private companies had the option to withdraw from the project, and the withdrawing party would be financially penalised. It was understood that entry into one phase did not involve a commitment to the next. The completion of Phase Two was critical: it would be the point of no return, for if the Governments and the two companies were still of one mind, the project would move to Phase Three – the start of the main Tunnel works. This would trigger the signing of a third agreement covering the period from early 1975 to the completion of construction in 1980.

Additional studies were to be commissioned by the Department of the Environment – one being a cost-benefit analysis to discover whether the Tunnel would be cheaper and a 'better buy' than the expansion and development of existing methods of crossing the Channel; and the other an assessment of its impact on the economic and social aspects of south-east Kent.

Leo d'Erlanger regarded the signing of the first agreement as the most significant step of the century towards the Tunnel and, on 1st December, 1972, proposed to the Company's shareholders a new rights issue of 464,000 shares to cover Channel Tunnel Investments' increased equity holding in the British Channel Tunnel Company. He concluded by saying that this should be the last call upon shareholders for further finance. The subscription of the new shares brought the Company's holdings in the British Channel Tunnel Company to 25%, and with that its pioneering role had been accomplished. The issue was encouragingly over-subscribed.

John Peyton had run into trouble earlier in October by informing the House that interim results of the studies, as was the case with all previous reports, would not be made public, but the Agreements would be lodged in the House of Commons Library. This fuelled growing criticism of the Government's apparent unwillingness to release information on the subject.

Successive Governments had given their assurance that the final decision on a Channel Tunnel would rest with Parliament. John Peyton rightly concluded that a premature public release of the incomplete studies would pre-empt the decision of the House – a fatal move for any Minister with aspirations to continue in Government – and could put the entire project in jeopardy. The schedule was extremely tight, and momentum would be lost if delay and indecision supervened. The completed studies would be followed by a White Paper, a Commons Debate and the presentation to the House of a short Money Bill. If Parliament approved, the project would move into Phase Two, to be ratified by the signing of Agreement No. 2 and an Anglo/French Treaty no later than 31st July, 1973.

Although the Parliamentary procedural aspects of the programme had been sanctioned by political judgement, the Minister decided that a Green Paper would be appropriate to ventilate the subject.

The Green Paper, entitled 'The Channel Tunnel Project'[46], was duly published on 31st March, 1973. It aired problems and procedures, and marshalled all known facts at the time, concluding with the statement:

> The Government welcomes informed consideration of the project by all who have an interest in it. A decision whether or not to proceed with the Channel tunnel is for this country an historic one. The Government wish the decision to be taken in the light of the fullest possible discussion.

The timing of the Paper, necessary as it may have been to John Peyton, was unfortunate. It did nothing to assuage the rising clamour for information. Because the final results of the studies were still three months away, the Green Paper could not include those facts that would enable 'informed consideration' to take place. Its publication was therefore something of a non-event and even intimidated some of the project's more staunch supporters.

The *Guardian* leader of 22nd March, 1973, under the title: 'Honest Doubt about the Tunnel', said:

> The Government does not yet seem to be certain whether a Channel Tunnel is desirable or not. Mr. Peyton commended the scheme yesterday with the unmistakable diffidence of a Minister who has not yet got his project past the Treasury. There will have to be a decision by July 31st as to whether the next phase of the undertaking – the first which involves actually digging – should go ahead or not . . .

The Times leader of the same date, headlined 'Very Probably a Mistake', said:

---

[46] Command 5256 – HMSO 1973.

The Channel Tunnel shows every sign of being the next large and costly public work to be foisted by the Government on the country without proper debate. Neither Parliament nor public have ever had adequate information on which to base an informed opinion of the benefit or otherwise of the tunnel to the country, or its effect on the lives of millions of people. The only previous official document, the 1963 White Paper, was totally inadequate for this purpose for reasons already rehearsed in these columns; and yesterday's Green Paper takes us little further . . .

This comment was notably restrained, for *The Times,* ever since the early 1960s, had consistently censured the project (as, indeed, it had in the 1880s), choosing to publish the more wild and woolly misconceptions of the day.

The editorial comment of the *Daily Mail* of 22nd March, 1973, succinctly summed up the situation under the heading of 'Slow Coach to Calais'. It said:

There must be a decision on the Channel Tunnel by this summer. Then, says today's Government Green Paper, 'full-scale boring work can begin'.

We have news for the Government. Full-scale boring work has been going on for 170 years, ever since Napoleon said 'Oui' to the idea.

Survey follows survey. Green Paper piles on White Paper. Yet still we are little the wiser about the commercial prospect of the Chunnel, about its likely effect on life in the South-East and about how to link the new quick route to the Continent with the Midlands and the North.

This latest 'further final' inquiry started exactly two years ago tomorrow. As the *Daily Mail* said then:

'*For heaven's sake, if the Channel Tunnel is a paying proposition, let's dig a hole now. And if it isn't, let's dig another hole and bury the whole idea.*'

Ten years earlier, almost the entire British Press had backed the project – and if it now seemed testy and hypercritical, there was good reason. It had been studied and talked about for far too long.

The House of Lords' debate on the Government's Green Paper was opened by Lord Orr-Ewing on 2nd May, 1973. He favoured the project, believing it to be 'a coherent, rational and viable plan'. Lord Harcourt gave the news that information resulting from the studies was beginning to filter through. Only that morning he had seen on the wall of the offices of the Project Management a copy of *The Graphic* of 24th June, 1879. The headline 'Proposed Channel Tunnel' was followed by a thought which he believed might be worth taking to heart:

The gigantic and long talked of scheme is one which must be admitted at least to be a splendid evidence of the determination and perseverance with which obstacles and impediments are met by the scientific men of the present age. Progress, though slow, is most encouraging.

'That', said Lord Harcourt, 'was ninety-seven years ago. But we are making progress and we are being encouraged.'

He continued by giving the Lords some idea of the present-day cost of the project. The cost in January 1973 terms was £470 million and on completion by 1980 it could be assumed to be between £820 and £830 million.

The debate lasted over four hours and observers judged that if a vote had been taken it would have favoured the Tunnel. Reactions, however, were shrill: 'Fears grow as Chunnel costs soar', headlined the *Folkestone Herald*. 'Chunnel far, far worse than Concorde . . . it would be a monument to the day when the British taxpayer put his hand in his pocket to improve a depressed area of France', said Keith Wickenden.

The assumed mantle of champion of the British taxpayer ill-suited Wickenden, who had succeeded his brother as Chairman of European Ferries. Most of his new vessels had been built in foreign shipyards and the special tax structure of shipping companies brought little by way of company tax to the Exchequer.

Concorde after its first landing at Filton, Bristol.

There was no denying that the public was confused with regard to the 'new costs', accustomed as they were to the ever-increasing charge on the taxpayer of that other Anglo/French venture, Concorde; but this was an ill-conceived comparison. Unlike Concorde, the construction of the Tunnel would not be a charge on the taxpayer. It would be built with private capital by proven engineering competence; whereas Concorde was handicapped by all the inherent problems and difficulties of innovation during construction.

Though the opponents of the project understood perfectly well, they were quick to exploit the lack of understanding elsewhere. The 'present-day' cost of the Tunnel of £470 million' simply represented the total costs if it was started on one day and finished the next. However, all was explained as, in response to the public concern over the lack of information, the British Channel Tunnel Company, on 15th May, 1973, pre-released a brief summary of the completed studies[47] that were being prepared for submission to the Governments.

> In summary the results of the economic, technical and financial studies show that there is a sound basis for proceeding with the construction of the Channel Tunnel. The British contribution to the cost of the tunnel at today's prices will be approximately £234 million, the French Company being responsible for raising an equal amount. After making provision for interest during construction and for inflation between now and the opening of the tunnel in 1980, the final cost is expected to be between £410 million and £425 million on each side of the Channel, a total of between £820 million and £850 million.
>
> The total operating profit (after deduction of £17 million

---

[47] See Appendix III.

operating costs) is expected to be between £90 million and £100 million in 1981, the first full year of operation. In the same year it is expected that some 9 million passengers without cars, six million passengers with cars and four and a half million tons of freight will pass through the Tunnel.

The Department of the Environment published 'The Channel Tunnel – its economic and social impact on Kent' on 21st May. Independently commissioned by the DoE and Kent County Council, in association with the Boroughs of Dover and Folkestone, the survey had been carried out by Economic Consultants Ltd., with the remit to study what the effects would be on the area if a Tunnel were built and what might happen if it were not. It was recognised that the Tunnel and related facilities would create a number of new jobs, but not necessarily suitable for workers made redundant from port or shipping employment. Suitable alternative employment would probably be found, although it was accepted that there could be a short-term adjustment problem.

Following the British Channel Tunnel Company's earlier statement of 15th May, the fuller report of the two Tunnel companies, 'The Channel Tunnel – Economic & Financial Studies[48], was published on 9th June, 1973.

The project once more reaffirmed its robust financial validity. In 1981 – its first full year of operation – after debt servicing and operating costs, there would be a net profit of £26 million, increasing to £163 million by 1990. The repayment of interest and principal would be the first charge on net Tunnel earnings and the total Government guaranteed debt would be completely repaid within 25 years. At the end of 50 years, the private companies' rights would cease and the Governments would be entitled to reap all the rewards.

The formula for equity shareholders presaged some tough bargaining. It would have to be sufficiently attractive to raise the capital; at the same time, neither Government would condone a formula which would bring excessive returns to the shareholders. The equity – classical risk capital in every sense – was to be a minimum of 10% and would be raised by the two Tunnel companies. The balance of the finance would not be a charge on the taxpayer. It would be raised by the two companies on the international money markets in the form of bond issues, which would carry a joint Government guarantee. The studies demonstrated that there would be little or no risk that the Governments would ever be called upon to honour their guarantees. The traffic and revenue studies had been based on a fare structure between 10% and 20% lower than fares charged on

---

[48] See Appendix III.

the short cross-Channel routes. Assuming the Tunnel was already in operation in 1973, the current ferry tariff for a medium-sized car with driver and two passengers was £21.60 – by Tunnel it would have been as low as £17.28.

The Tunnel would be a viable proposition on a terminal to terminal basis, but to maximise its full potential it would be essential to construct a new rail track from London to the Cheriton Terminal as the existing tracks from Folkestone to London could not accommodate the growth of the generated passenger and freight traffic. Whereas British Rail rolling-stock already travelled on Continental rail systems – as the track gauge[49] is the same – the loading gauge (or dimensions) of Continental rolling-stock, is both higher and wider. BR's preferred site for the London terminal was at White City and its estimate for the terminal and the new high-speed line was approximately £120 million.

The potential growth in Continental passengers and freight held forth the promise of a rich harvest in the future for the new 25 k.v. overhead transmission high-speed route. BR had been unable to develop cheap freight services because of geographic constraints – the longest freight haul in Britain being some 300 miles. It is cheaper to send freight by road up to distances of around 300 miles, but with the Tunnel link BR would extend its route potential by thousands of miles and break through this 'tariff barrier'. Freight rates become progressively cheaper the further it is carried and the overall effect would be a dramatic diversion of freight from road to rail.

With the introduction in the mid-1980s of BR's APT[50] the passenger journey time from London to Paris and London to Brussels would be reduced to 2 hours 30 minutes and 2 hours 15 minutes respectively, with corresponding time savings for other destinations. It was estimated by the Civil Aviation Authority that the effect of BR's city-centre to city-centre journey times would reduce aircraft movements in the London area by some 20,000 in 1985.

A model of BR's APT.

Following on the heels of the British Channel Tunnel Company's economic and financial report, the DoE's cost benefit study[51], which had been carried out by Cooper Lybrand & Associates, was published on 13th June. Comparing the cost of the development and expansion of existing means of cross-Channel transport to meet the freight and traffic demands of the future with the cost of building a Tunnel, the report concluded that, although the Tunnel initially required a higher capital investment, in the long-term its low maintenance and operating cost conclusively

---

[49] The distance between the two railway lines.
[50] See Appendix III.
[51] *The Channel Tunnel: A United Kingdom Transport Cost and Benefit Study,* HMSO.

demonstrated that for the nation it was the 'best buy', being cheaper by some £290 million (in 1973 terms).

This last study completed the picture. During the past two years the project had been 'out through the wringer' and was once more shown to be financially viable and the cheapest and most efficient means of deploying the nations' resources. Earlier studies had produced similar results – but this time there was no room for doubt.

On publication of the 1973 studies, the Channel Tunnel Opposition Association erupted in a storm of protest concentrating on the effects that the siting of the terminal at Cheriton would have on the environment. Yet the Centre for Environmental Studies, in the person of Professor David Eversley, took a very different, indeed positive view:

> It doesn't make any kind of sense for the Government to judge the tunnel on the basis of whether it will be financially viable. It will be the greatest contribution to the environment of the south east and should be built regardless of cost – otherwise Kent will have to be concreted over to take the traffic of the future.

Even so, fears on either count were none the less real. 'Build it anywhere else but here!' became the rallying-cry and on an over-populated island, anywhere else is inevitably in someone else's back garden! Concern for the environment has become a laudable pursuit and protests from conservationists are expected whenever development of any kind is proposed. Invariably the site is the habitat of some rare flora or fauna. Most people had never heard of the Brent goose until it was proposed to site London's third airport at Maplin. Similarly, it was revealed that the rare Spider Orchid flourished in the area chosen for the Cheriton Terminal. Such habitats, until threatened, are always jealously guarded for fear that the uninitiated might unthinkingly disturb or even destroy them. Man's uncontrolled progress in the past has much to answer for and no one should condone the wilful and reckless destruction of nature's gifts, but even in the carefully controlled planning of our future environment, it should be remembered that life is also about people.

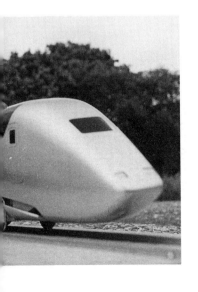

RTZ's Project Managers could not be faulted on their facts and figures – even the shipping companies conceded that the traffic and freight forecasts were a fair assessment of future trends. They now directed their energies towards creating a wider understanding and faith in the study results.

DEBATE AND LEGISLATION

Just over a month later, on 15th June, the studies were the topic of a five-hour Adjournment Debate in Parliament. John Peyton emphasised that the Government did not intend either to announce a decision[52], or to ask the House to make one by implication; more time was needed by Government, Parliament and the public to consider the information which was now available. He described the studies as showing the project to be technically feasible and financially sound. The study of the economic and social impact on Kent had reached a reasonably reassuring conclusion. He continued by outlining the legislative procedures if the project were to be implemented. It would be necessary to introduce a Short Money Bill and thereafter a major Hybrid Bill to ratify construction and the main financial and operating arrangements. He concluded by reminding the House that the choice before them was either to go ahead, or to abandon the project entirely. 'We cannot opt for cold storage', he said. Emphasising his earlier observation that the traffic through Kent would continue to increase, he concluded: '. . . it will not go away. It must, in the absence of some very Draconian restrictive policies, be catered for . . .'

The principal spokesman for the Opposition, Anthony Crosland, Shadow Secretary of State for the Environment, welcomed the fact that the Minister seemed to have no wish to railroad a decision and he proposed that the House should take advantage of the terms of Agreement No. 1 and postpone its decision until 15th November. Crosland admitted that he had been sceptical, just as he was towards London's proposed third airport at Maplin, Concorde, and other grandiose projects. In the light of current evidence, however, he had moved to a more favourable position on the Tunnel: 'I am perhaps half way along the road to Damascus, but I have yet to see the blinding light.' He now believed that there was a strong *prima facie* case for the project's commercial viability; he found the conclusions of the Cost Benefit Study difficult to rebut; and providing there were no adverse regional and environmental effects and the railway investment was realistic, he found the case for the Tunnel persuasive. However, it was unlikely that those aspects, which were yet to be proved, could be properly considered by the end of July.

To complete within six weeks an agreement with the Governments on the shareholders' formula, the publishing of a White Paper, the passing of a Short Money Bill, Agreement No. 2 and a duly signed Anglo/French Treaty, was an agenda too daunting even to contemplate.

Although the House had been impressed with the thoroughness of the Project Management's report, the overall inclination erred towards caution. It would be impolitic of John Peyton not to take note of the views of the House, as it would lend

---

[52] The Houses of Parliament traditionally rose for the Summer Recess at the end of July and resumed in October.

substance to the already voiced criticism that the Government was trying to railroad the project, or that, following the Heath-Pompidou meeting on May 21st, it was already a *fait accompli*. The Government, too, were perhaps conscious that their back-benchers were restless, following their rebuff on 13th June of the Third Reading of the Government's Maplin Sands, Third London Airport Bill.

The final report of the two years' studies, including engineering and technical appendices, and totalling some 32 volumes, was presented by the Project Management to the Government at the end of June 1973. The task had been completed on time and under budget. The technical design and concept of the Tunnel did not differ in any great detail from the Channel Tunnel Study Group's earlier design[53].

There was a certain unanimity in the two views which were expressed from both sides of Parliament. Lord Shackleton[54], Chairman of RTZ-DE, was Leader of the Opposition in the House of Lords; and, as guest speaker at the Westminster Chamber of Commerce at the end of June, he said knowledgeably: 'It is understood that any serious hold-up could add several million pounds to the bill.' Prime Minister, Edward Heath, in his 'end of term' speech to the 1922 'back bench' Committee on 19th July, confirmed his belief: 'It is economically right.'

The day before Parliament rose for the Summer Recess, John Peyton made the following statement to the House:

> Financial negotiations between my French colleague and myself and the private interests are still continuing but they have not yet been concluded. This is necessary before the Government can take their decision on the project . . . The Government's decision, when it is taken, will be announced in the form of a White Paper. If it is affirmative, draft clauses of a Money Bill which we would propose to introduce into the House immediately we resume in October would be included in the White Paper. There is, of course, no question of signing any agreement or Anglo-French Treaty before this Bill is passed . . . We would wish to provide the fullest possible opportunity for it to be considered, by the public as well as MPs, before we resume in October.

This delay of at least three months was part of an all too-familiar pattern. Was history about to repeat itself yet again?

Alistair Frame[55] (later knighted) was Managing Director of Rio Tinto

Sir Alistair Frame, Managing Director of RTZ-DE.

---

[53] See Appendix III.

[54] Son of the famous explorer, Sir Ernest Shackleton.

[55] A Scotsman born in Glasgow in 1929. Prior to joining RTZ-DE, he had been Director of Reactor Design in the UK Atomic Energy Authority, and a member of the Managing Boards of both Reactor and Harwell Research Groups of the AEA.

Zinc Development Enterprises, which had established an enviable record of achievement in the management of major projects throughout the world. As project manager to the British Channel Tunnel Company, he was endowed with all the qualities and experience that the project would demand. He had already demonstrated the worth and capability of RTZ-DE's appointment, not only with the completion of Phase One within time and budget, but with the disclosure of an unexpected bonus which more than paid for the cost of the two years' study. His team had also proposed a new route for the Tunnel on the British side. This new line in the vicinity of the old 1880 workings at Shakespeare Cliff would shorten the overall length by two miles, a saving of £7 million in construction costs and a subsequent reduction in annual operating costs of £300,000.

Alistair Frame's French counterpart was Jean Gabriel[56], *Ingénieur en Chef des Ponts et Chaussées* and Project Director of *Société Française du Tunnel sous la Manche.*

During the two-year study period of Phase One, these two engineers had forged a firm rapport, formulating an integrated working relationship which would be essential for the co-ordination of the task that lay ahead.

Undeterred by Peyton's postponement of a decision until the late Autumn, and in anticipation of Parliament taking a favourable view, Alistair Frame and Jean Gabriel completed preliminary plans to make an immediate start in November. It was announced during the second week in August that they had nominated the contractors who would undertake the £30 million Phase Two Tunnel works[57].

Meanwhile, the Tunnel companies were engaged in meetings with Government officials on both sides of the Channel in an endeavour to conclude the critical matter of the shareholders' formula. The officials, no doubt encouraged by the financial viability demonstrated by the economic and financial studies, were determined to strike a hard bargain, as the balance of the net annual profits after the payment of interest on the equity share capital would be equally divided between the two Governments.

The interests of both public and private sectors were eventually reconciled and agreement was reached at a final meeting

Jean Gabriel, Project Director of Société Françoise du Tunnel Sous la Manche.

---

[56] Born at Narbonne in 1921, Jean Gabriel was a product of the French elite academy – *l'École Polytechnique.* During his career he became eminently qualified in the technical, economic and financial implications of major transportation schemes. One-time Chief of the Transportation Road Division of the World Bank, he was also responsible, as Port Operations Director, for major works on the French Naval Base at Mers el Kebir and the port and airport of Algiers.

[57] British consortium: *Cross Channel Contractors* comprising Taylor Woodrow Construction, Guy Atkinson, Balfour Beatty and Edmund Nuttall. French consortium: *l'Entreprise Industrielle* comprising *Société Anonyme des Entreprises Leon Ballot, l'Entreprise Capag-Cetra, l'Entreprise Truchetet-Tansini, l'Enterprise Quillery-Saint-Maur, Heitkamp GmbH (RFA) and Trapp et Cie GmbH (RFA).*

in Paris on 5th September, 1973, between John Peyton and Pierre Billecocq, deputy to Yves Guena, the French Minister of Transport.

During the following week the British Government announced its official blessing for the project in a White Paper[58], almost ten years to the day from publication of the 1963 White Paper: *Proposals for a Fixed Channel Link.*

At the press conference on 12th September, John Peyton, flanked by Richard Marsh for British Rail and Lord Harcourt and Alistair Frame for the British Channel Tunnel Company, announced that the British and French Governments had decided, subject to legislation, that it would be in their joint national interest to build the Channel Tunnel with a new high-speed rail link to London.

The financial arrangements would be a unique partnership of public and private enterprise on an international basis. The terms under which the private interests would receive a fair reward for the risks to be taken were understandably complex, and allowances would have to be made for adjustments in accord with any significant changes in future market conditions. During construction the risk capital would receive a flat return of 7% per annum. It had, however, been agreed with the two companies that in the present climate of world-wide inflationary trends, and the UK Bank Rate having reached the unprecedented level of 11.5%, that it would be prudent to take no final decision until the start of Phase Three in 1975, when the bulk of the risk capital would be raised. In the light of prevailing market conditions, payments would subsequently be made to the companies under a formula based on gross and net revenues, coupled with a flat percentage return.

The projected revenue forecast promised a lucrative return to the Governments and an equitable return to the private shareholders. Based on the lower potential growth forecast of Tunnel traffic, and after deducting operating costs, bond servicing, and interest payments, the Governments would share equally £5 million from net profits in the first full year of operation in 1981; this would rise to £120 million a year by 1990; and £375 million in the year 2000. The equity shareholders would correspondingly receive interest payments of £22 million in 1981, rising to £40 million a year by 1990 and £75 million in the year 2000.

Richard Marsh, who described himself as having always been 'a confirmed Tunneller', was obviously delighted with the Government's approval of British Rail's plans for a new high-speed Berne gauge line from the terminal at Cheriton to the White City Terminal, London. 'This is', he said 'one of those decisions by Government which is perfection itself.'

The tide in the fortunes of the Tunnel had once more turned

---

[58] *The Channel Tunnel* (Command 5430) HMSO.

in its favour, and this time the groundswell of declared official intent and goodwill prevailing on both sides of the Channel appeared to embody sufficient surge to overcome all future obstacles.

It was at this juncture that there was a dramatic 'last ditch' display of competitive spirit from the UK Chamber of Shipping. It claimed that the Tunnel's traffic consultants' calculations of the Tunnel's share of accompanied cross-Channel car traffic was 'completely unrealistic', and argued that the maximum expectation would be at least 40% less. Furthermore, the ferry operators could under-cut Tunnel tariffs and so compel the Governments to subsidise the Tunnel in the early critical years. It did, however, admit that the sea-ferries would eventually be forced out by such subsidised, unfair competition.

A more 'Luddite' formula for committing commercial suicide would be difficult to imagine – but the Project Management's response was more logical.

In the first year of Tunnel operations the forecasted Tunnel revenues were based on a fare structure that would be 18.5% less than the fares on the short cross-Channel routes. Therefore, the sea-ferry operators would have to reduce their fares by 18.5% in real terms to be on par with Tunnel tariffs in 1980; if they were to reduce fares by, say, a further 30%, the Project Management concluded that even this extreme competition would not materially affect Tunnel revenues. However, the consequences of reducing sea-ferry fares by over 40% in real terms and still maintaining a reasonable return on capital would mean a severe reduction of their fleet, operating a reduced service with no capacity to contend with seasonal peaks.

If this margin of flexibility existed in the ferry operators' tariff structures it confirms that this narrow stretch of water is the most expensive ferry journey in the world.

The Government's approval for the high-speed rail track between White City and the Cheriton Terminal in Folkestone provoked further protest. The Channel Tunnel Opposition Association had been campaigning against the siting of the Tunnel terminal at Cheriton; it was now determined to contest the proposed 75 miles of new track every inch of the way.

The Department of the Environment had called upon British Rail to conduct a public consultation programme akin to the procedures in practice for a new motorway route, in order to conciliate the objectors in Kent. This was unfortunate for BR who, apart from being totally ill-equipped to undertake such consultations, would be entering a public arena where opposition forces, particularly in Kent, had enjoyed an unbridled run for some months.

This unfortunate situation arose from Government officials advising the Tunnel Company earlier in the year to adopt a low profile, and not to further inflame the position by responding to criticism of the project. The company reluctantly followed this ill-conceived advice, unwisely as it proved, as it was subsequently privately and unfairly admonished by John Peyton. The gist of what he told members, forcibly and in very unparliamentary language, was: 'If you want the Tunnel, go out and get it!'

Parliament was prorogued on 24th October, and the following day, just before the Secretary of State for the Environment, Geoffrey Rippon, rose to move the motion that the House should approve the White Paper, the Speaker read out an amendment to the motion in the name of the Leader of the Opposition, Harold Wilson. It was a mean-spirited little amendment. While it did not oppose the Tunnel in principle, it withheld approval of the scheme in its present form, and demanded an independent enquiry into alternative transport strategies, including a 'rail only' tunnel[59].

During the debate that followed, Eric Ogden, Labour Member for Liverpool, West Derby, in the course of an impassioned point by point rebuttal of the opponents to the project, made a swingeing attack on his own Party for entering the amendment. He was in no doubt that Labour Government policy from 1964 to 1970 favoured the scheme and was proud of what Labour Ministers and the Labour Government had done to carry it forward. There might be points of difference with the present scheme – but he argued that legislation would provide ample time to sort out any difficulties and it was in these circumstances that he could not reconcile himself to Wilson's amendment. 'It is a wholly negative, pessimistic, horrible little amendment, and I want no part of it. I want the Tunnel. I want the Government, who are only occasionally right, to get the proposal through!' He also made generous reference to Albert Costain, Conservative Member for Folkestone, who was now a Junior Minister in the Department of the Environment and, as such, unable to participate in Parliamentary debate. His constituents had given him a rough time, accusing him of failure to represent their views to the House: 'Mr. Costain has been criticised for keeping quiet in public. He has to, and that was a most unfair criticism. Privately – ye gods! – I believe he has been making the right observations in the right place!' His constituents failed to realise that, as Parliamentary Private Secretary to Geoffrey Rippon, he was in a particularly advantageous position to gain the ear of the Minister.

Eric Ogden was one of the staunchest supporters of the Tunnel in the House. His vision was that the Tunnel would revitalise Liverpool Docks which would become the Western

---

[59] Rail passengers and freight only – no road vehicles.

Gateway to Europe with transatlantic cargoes being off-loaded at Liverpool on to trains which would carry them direct to their Continental destinations.

After nearly six hours of debate, the House divided and 436 Members recorded their votes – Eric Ogden, defying a two-line Whip, abstained. The result: 243 voted for and 187 against – a Government majority of 56.

In the following days the Channel Tunnel (Initial Finance) Bill was passed as a formality and received the Royal Assent on 13th November, 1973.

The official baptism – the signing of Agreement No. 2 between the two Governments and the Tunnel Companies – was witnessed and recorded by an international press corps and a battery of television cameras at Lancaster House on 17th November, 1973. The principal signatories were John Peyton, Minister of Transport Industries; Pierre Billecocq, the French Minister of Transport; Lord Harcourt, Chairman of the British Channel Tunnel Company; and General Philippe Maurin, President of the *Société Française du Tunnel Sous la Manche.*

A smiling John Peyton described the signing as 'an historic occasion' which came 171 years after the idea of a Tunnel was first mooted. An equally serene Pierre Billecocq, beginning his speech in French and changing midway to English, said the ceremony meant a physical link which would 'contribute immensely to strengthening the personal ties between the people of Britain and Europe'. He concluded with the quip 'It shall be easier for you to come and pick our roses of Picardy and for us it shall no more be a long way to Tipperary'.

Later that day the signed agreement was taken to Chequers where President Pompidou was weekending for talks on the Common Market and the Middle East with Prime Minister Heath. The document was appended to an Anglo/French Treaty which was signed in an atmosphere of great cordiality by Sir Alec Douglas-Home, the Foreign Secretary, and the French Foreign Minister, Michel Jobert. The Treaty would not come into force immediately, but would have to be ratified by 1st January, 1975. The Agreement authorised the construction of initial tunnel works and two kilometres of pilot tunnel from both the French and British coasts; it was to be completed by mid-1975 at a cost of £30 million, shared equally between Britain and France, £8 million being subscribed by the two Tunnel Companies and the balance of £22 million by way of a joint Government loan to the two companies.

CHAPTER
FOUR

THE
SECOND
ATTEMPT

1974 – 1975

The two Tunnel companies wasted no time. On 19th November Lord Harcourt and Alistair Frame on behalf of the British Channel Tunnel Company, signed a £6 million contract for the preparatory trial works with David Balfour, Chairman of Cross-Channel Contractors. A similar ceremony was conducted by General Philippe Maurin on behalf of *Société Française du Tunnel Sous la Manche* with *l'Entreprise Industrielle*. The following day, after lying fallow for some 90 years, the sites at Sangatte and Shakespeare Cliff began to bustle with activity as heavy equipment was moved in and bulldozers began to clear the ground.

The first task at the top of Shakespeare Cliff was to drive an access road tunnel 287 metres (314 yards) in length, which would be angled through the cliff, and pass over the railway tunnel down to the foreshore site, adjacent to the Folkestone-to-Dover main railway line. A second tunnel, 451 metres (493 yards) long, 6.5 metres (21.3 feet) in diameter, would be driven from the foreshore with a gradient of 1:6 to the starting point beneath the sea-bed of the two kilometres (1.2 miles) of pilot tunnel. These access tunnels were by definition the means whereby the boring machine could be located at the correct operating depth and for the removal of the spoil.

Work on the pilot tunnel – 5.27 metres (17.3 feet) in external diameter, lined with reinforced pre-cast concrete segments 36 centimetres (14 inches) thick – was expected to start in September 1974 with the delivery to the site of the full-face Priestley rotary tunnelling machine. The 95,000 m^3 spoil from Phase Two works would be deposited inland of Shakespeare Cliff where it would be used to level the ground for possible agricultural use. Spoil from

Showing 1974 pilot tunnel boring towards France on left; and access tunnel and boring operations on right.

Phase Three would be 3.5 million m^3; this would be transported by rail from the Tunnel workings direct to the terminal site at Cheriton where it would be used for levelling and subsequent extensive landscaping of the area. Aesthetic appeal was one of the principal features of the design of the terminal and it would be effectively screened with grass-covered, tree-lined embankments.

On the site at Sangatte only one access tunnel, some 214 metres (234 yards) long, 7.8. metres (25.6 feet) in external diameter, with a gradient of 1:9 would be necessary to descend to the level of the pilot tunnel beneath the sea-bed. The French Company had chosen the American-designed Robbins full-face rotary tunnelling

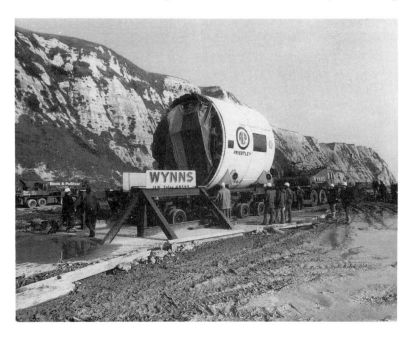

The cutting head of the Priestley boring machine arrives on site at Shakespeare Cliff.

machine for their pilot tunnel and it was expected that a start on the two-kilometre-length would be made in October 1974. The type of tunnel lining would differ from that used on the British side and would be ribbed cast-iron segments. Whereas pre-cast concrete tunnel linking segments are cheaper but more time-consuming to erect, cast-iron segments are more expensive, but quicker and easier to erect. The operational experience gained in Phase Two, together with the relative cost and handling properties of these two methods, would be carefully evaluated before deciding on the type of lining

The Robbins boring machine chosen by the French Channel Tunnel Company.

that would be used throughout the main Tunnel. However, cast-iron segments would always be used where the Tunnel passed through suspect ground and at the corners of the cross-adits which would occur every 250 metres (273 yards).

The Channel Tunnel had always been a hostage to fortune and now, when all the portents seemed set fair, political storm clouds precipitated by world events began to cast a shadow of uncertainty. In October 1973 war erupted between Israel, Egypt and Syria, and with it came the oil crisis. The Arab States held the principal world users to ransom by reducing supplies, accompanied by a three-fold price increase. Britain, riven by escalating retail prices and wage demands, was dealt a further blow by a head-on clash between the Government and the mineworkers. The Government refused the NUM's wage demands which resulted in a ban on overtime and the threat of strike action which, with the oil crisis, created a grave energy problem. Edward Heath imposed a State of Emergency and, to conserve power, declared a three-day working week. This worsening state of affairs was brought to a head with the Government going to the country on the issue of 'who rules the country, the elected Government, or the Unions?' and a General Election was set for 28th February, 1974.

During the three-week run-up to the Election, references to the Tunnel were as discouraging as they were frequent. Harold Wilson proclaimed that if he were returned to power he would view the project 'with a jaundiced eye' and Jo Grimond, Leader of the Liberal Party, dismissed it as 'a Government extravagance which no one would ever use'.

The outcome of the Election was stalemate. Neither party received a clear mandate, and each was denied an overall working majority. Heath's Government resigned, and Harold Wilson became Prime Minister and formed a minority Government. This political uncertainty placed in jeopardy the shift of Tunnel construction from Phase Two to Phase Three planned for 1975. The project's sensitivity to political oscillation was to be further exposed by the death of Georges Pompidou. After a protracted illness courageously borne for over a year, he died suddenly on the night of 2nd April, 1974.

The following day, while the French nation mourned the passing of their President, Anthony Crosland, now Secretary of State for Transport, swept speculation and doubt aside. Having said earlier that he was part-way along the road to Damascus, he now saw the light. He announced to the House that Phase Two would continue to completion and that he would be re-introducing the Tunnel's Hybrid Bill. However, although the current programme provided for a reassessment of previous studies, he intended, as a matter of urgency, to obtain an independent appreciation with special reference to maximising the use of through-rail services. He also reaffirmed his rejection of the rail-only Tunnel concept. The Tunnel's Hybrid Bill, whose enactment was essential for Agreement No. 3 and the ratification of the Anglo/French Treaty, was re-introduced into the Parliamentary programme on 10th April, 1974; it completed its Second Reading on 30th April with a comfortable majority of 287 votes to 63. On the very same day, Fred Mulley, Minister for Transport, announced that Sir Alec Cairncross, formally

The lower foreshore site of Channel Tunnel Phase Two construction works at the foot of Shakespeare Cliff, Dover.

These five tunnels span a period of more than 130 years. The tunnel leading to the viaduct and the adjacent gallery under the cliff were built in 1974. The twin railway tunnels on the extreme left date from 1843-44. The centre tunnel provides access to boring operations; and the tunnel up the ramp provides access to the upper working site. In the foreground is the shaft of the 1880-82 pilot tunnel. Descending vertically for some 160 feet the shaft connects with the 2,024-yard, seven-foot-diameter tunnel driven by Captain English's boring machine along a route extending towards Dover Harbour. The shaft and tunnel, which was flooded, had to be drained and plugged with concrete at the point where the present pilot tunnel actually bisected Beaumont's tunnel.

head of the Government's economic services, was appointed Chairman of the Channel Tunnel Advisory Group. The remit of the Group was to produce the independent assessment that Crosland had announced, before final commitment to Phase Three.

Even allowing for the precarious position of the Wilson Government, coupled with the possibility of post-Pompidou hesitation in France, it now seemed that the Channel Tunnel would be finished and operating no later than 1980.

Meanwhile, Phase Two works at the Shakespeare Cliff site were continuing on schedule. The new Tunnel route would cut through the 1880 Beaumont tunnel almost at right-angles beneath the sea bed. To ensure that with the passage of years it had not flooded, Cross-Channel Contractors uncapped and cleared the main shaft to the old workings. Unlined, the tunnel was dry as the day it had been excavated nearly a century ago. If any further confirmation was needed, this proved the integrity and characteristics of the Lower Chalk.

With an overall majority of six, the Wilson Government had made a shaky alliance with the Liberal Party. After eight months Wilson found the position intolerable and another General Election was held. The Labour Government was returned, on 10th October, 1974, with a more workable majority.

The long-held vision by BR and SNCF – of closing that 21-mile gap – had waned with BR in recent years. When Dr. Michael Bonavia[60] was appointed Director of BR's Channel Tunnel Project Department in the 1970s (ably assisted by David Williams) he found little support for his small overworked staff – everyone was too busy running a railway!

In September 1973 John Peyton had given BR the go-ahead for a dedicated high-speed line from London to the Tunnel terminal, with plans and costs incorporated in a BR Private Bill by November 1974. A daunting task but BR met the deadline – at a cost!

The original cost estimate in 1972 was £120 million which had escalated to £375 million. Asked to design a high quality rail link BR's designers and engineers had obliged. They proposed a tunnel from London to Croydon with a three-lane maintenance road running alongside throughout its length!

Anthony Crosland told the House on 26th November:

> It is out of the question that the Government should approve, or finance an investment of this magnitude. We must find some less expensive means of enabling the through-rail traffic, which forms so essential an aspect of the Tunnel project.

He charged BR with the urgent task of examining lower cost options; and, as it was now impossible to adhere to the time-table, he proposed to the French Government and the two Tunnel companies that the time-table be put back to allow a lower cost rail solution to be examined '. . . before we decide whether to build the

---

[60] Author of *The Channel Tunnel Story* (David & Charles, 1987).

Tunnel, or not'. He asked for early indications of their readiness to re-negotiate arrangements on this basis. In conclusion he said:

> As I have repeatedly told the House, the decision whether to proceed with Phase three and build the tunnel remains completely open, and the house will have the fullest opportunity for debate before this final decision is taken.

The reaction of the French Government to this domestic *contretemps* was quite straightforward. Again Crosland informed the House:

> . . . the French Government have now re-emphasised to Her Majesty's Government their intention to complete Phase two and the current economic studies with a view to signing Agreement three within the agreed time schedule.

Workings at Shakespeare Cliff. To the left is the access tunnel, entering the pilot tunnel. The pilot tunnel towards France continues in the foreground out of sight. The short section of pilot tunnel depicted would have continued towards the terminal site at Cheriton.

The British Government could not ratify the Anglo/French Treaty on 1st January, 1975, without the prior enactment of the Hybrid Bill and it was not possible for the Bill to obtain Royal Assent before that date. At best, a mutually agreed re-scheduling of the works programme, or at worst an agreed, temporary postponement of the project was the only solution for the Government, which would otherwise be in default of Agreements with the French Government and both Tunnel companies.

Prior to Anthony Crosland's statement it had been agreed in Cabinet that he should negotiate a year's delay. The French Government was approached and, on 9th December, the French Minister of Transport, Marcel Caraille, responded: 'The French commitment to the Tunnel remained steadfast; and discussions should take place with a view to revising the timetable.' Just over a week later, on 17th December, the French, ever confident, passed a Bill authorising the ratification of the Treaty.

1st January, 1975, came and went. On 2nd January, the Tunnel companies served formal notice on the Secretary of State for Transport that the British Government was in breach of its contractual obligations.

The Tunnel companies were obliged to take this step to protect the interests of their shareholders, as the prospectus which they had issued earlier to raise £8 million, was a legal document, describing in detail the timing and the purpose of this invitation to investors. However, on 9th January the Tunnel companies sent a letter to Crosland acknowledging the Government's dilemma; but, having no wish to conclude matters, they proposed a new timetable: the Tunnel bill to be re-introduced in the Autumn of 1975, enacted by mid-1976, with a commitment to start full-scale works by the end of 1976.

Meanwhile, in Sangatte, the Robbins boring machine was poised at the entrance of the French access tunnel, ready to descend to start boring operations. At Shakespeare Cliff, the Priestley machine was already in position and was in the process of completing commissioning trials.

On 20th January, 1975, the Priestley had already started full boring operations – but at 4.00 p.m. the news from Westminster filtered through and the machine was switched off. History had repeated itself. The Channel Tunnel had been unilaterally abandoned by the Wilson Government.

CHAPTER
FIVE

HISTORY
REPEATS
ITSELF

1975 – 1987

There are few projects against which there exists a deeper and more enduring prejudice than the construction of a railway tunnel between Dover and Calais.

Again and again it has been brought forward under powerful and influential sponsorship. Again and again it has been prevented. Governments of every hue, Prime Ministers of every calibre, have been found during successive generations inflexibly opposed to it. To those who have consistently favoured the idea this ponderous and overwhelming resistance has always seemed a mystery.

Winston Churchill, 1936

And mystery it continued to be. The Anglo/French Treaty had been ratified by the National Assembly on 17th December, 1974, the two Tunnel companies were willing to negotiate a new agreement and the French Minister of Transport, Marcel Caraille – whilst expressing the French Government's attachment to the project – had agreed to negotiate a new timetable.

There was no cause to blame BR for the delay, as alternative cheaper options for the high-speed track were available for consideration.

Anthony Crosland, having arrived at Damascus a convert, deeply regretted the Government's decision. He still firmly believed that he would see the Tunnel built in his lifetime.

Was it the Treasury mandarins? It was known that they had objected strongly to public funds being earmarked as a liability contingent to the Government's guarantee of the Tunnel bonds – but they had been overruled.

Was it the declining economy? In the context of the worldwide depression caused by the Middle East oil crisis, the Government could have given this for its reason for the unilateral abandonment of the project – but it chose not to.

However, Harold Wilson was poised to introduce government control of incomes and prices. Because it would be a first-time measure for a Socialist Government he wisely sought the reaction of the TUC. His emissary was Jack Jones, General Secretary of the Transport & General Workers' Union, who came back with a dusty answer. Yes, he could count on TUC support, but the price would be the cancellation of the three 'white elephants' in public sector spending: Concorde, London's third airport at Maplin, and the Channel Tunnel. It would cost more to cancel than to continue with Concorde; plans for Maplin were still in their infancy; so to appease the unions, the Channel Tunnel, although not funded with state finance, was offered up as a sacrifice. Harold Wilson subsequently introduced his incomes and prices policy, which he called the *Social Compact*, but within a week it had been dubbed the *Social Contract*.

Schematic showing honeycomb of tunnels built during the first and second attempts.

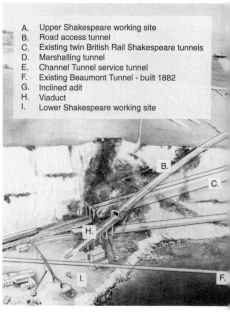

A. Upper Shakespeare working site
B. Road access tunnel
C. Existing twin British Rail Shakespeare tunnels
D. Marshalling tunnel
E. Channel Tunnel service tunnel
F. Existing Beaumont Tunnel - built 1882
G. Inclined adit
H. Viaduct
I. Lower Shakespeare working site

All of which was indicative that the Tunnel was still a political football, vulnerable to trade off in return for short-term gain elsewhere. Successive Governments had played this game, giving the decision to go ahead, only to lack the political will to see it through to completion. The result was not only repeated disillusionment but personal sacrifice. One in lighter vein is wryly recalled by Richard Marsh.

He was a frequent visitor to the French Embassy on social occasions, and early in January the Ambassador had asked for his reading of the present position on the Tunnel. Only two days before the project was cancelled, he told the Ambassador that in his opinion the situation was no more than a 'temporary hiccup'. On the following Friday, as he was passing through the Central Lobby in the House of Commons, he was hailed by Fred Mulley with the words: 'You can forget about your Channel Tunnel, it's been scrubbed!' Meantime, of course, the French Ambassador had informed Paris of his reading of the situation. The outcome, as Richard Marsh relates, was that 'I became less of a frequent visitor to the Embassy – which was a great pity as they served a very drinkable glass of claret!'

The sad process of winding up the project began, but Crosland assured the House that 'the studies, plans, and works will be preserved as far as practicable to avoid waste, should the Tunnel scheme be revived.'

The Department of Transport commissioned Cross-Channel Contractors to 'mothball' the Priestley machine, which meant the boring of some 250 metres of lined tunnel[61]. This exercise gave the Department valuable data on the performance of the scarcely-used machine and tunnel lining in chalk conditions.

The prudent French built a shed over the Robbins machine, which remained poised at the mouth of their access tunnel which they allowed to become flooded.

Compensation of some £8.5 million was paid to each Tunnel company. As was to be expected, the French Government paid its share grudgingly, but such was the nature of the original agreement. The assets of the two companies now became the property of the respective Departments of Transport. Channel Tunnel Investments, however, on behalf of the Channel Tunnel Study Group, withdrew its geological studies of 1958-59 upon which the major 1964-65 Geological and Geophysical Survey had been based. The reason for this was a passage in the White Paper which specified that when the Tunnel was completed, the Study Group would receive £3.7 million in compensation for its earlier endeavours and studies. Therefore, in principle, any future construction of the Tunnel could not make use of the 1964-65 geological information without reimbursing the Study Group.

A more unusual consequence of the abandonment was that the Department of Transport now owned a Public Company – the British Channel Tunnel Company and the Minister found himself its Chairman ex-officio. With a board of directors made up of his senior civil servants, the Minister would solemnly hold Annual General Meetings.

The final irony came with the publication of the Report of Sir Alec Cairncross's Channel Tunnel Advisory Group, which concluded in favour of the project – six months after abandonment!

After twenty years of frustrated endeavour, the most fervent supporters feared it was the last one would hear of the Channel Tunnel this side of the twenty-first century.

NEW PROPOSALS, NEW GOVERNMENT

Yet the vision of a Channel Tunnel had not vanished. After reaching such a disappointing climax for the second time in its history, the challenge remained.

Only two years later British and French Railways picked up the gauntlet, having concluded that perhaps the sheer magnitude of the scheme had been the architect of its demise. Might a Tunnel more

---

[61] The total length of the boring machine and its supporting 'train' for erecting the linings and spoil removal was 250 metres.

modest in concept and cost, be viable and more acceptable to Governments and the public?

The Tunnel Project Management had worked up several schemes more moderate in cost and scale, one of which was adopted, with its blessing, by BR and SNCF. It was a single-track bored tunnel and service tunnel, restricted to rail passenger and freight traffic only. The single main tunnel would be 6.02 metres (19 feet 9 inches) in diameter, with a pilot tunnel of 4.5 metres (14 feet 8 inches) in diameter. It would follow the same cross-Channel route as the 1974-75 scheme, for an estimated capital cost of £650 million, including the built-in capability for the addition of a second main tunnel at some future date to meet expanding traffic demands.

The effective traffic pattern of single-track tunnel operation would involve convoys of trains, spaced at regulated intervals travelling through the Tunnel in one direction for a set time, alternating with trains travelling in the reverse direction for a similar time span. It would operate on a two-phase, three-hour cycle: ten trains would be accepted through the Tunnel in the first one hour and twenty-five minutes, followed by a ten-minute break, before accepting ten trains from the reverse direction during the next hour and twenty-five minutes. The traffic pattern would be a mix of passenger and freight; and, allowing six hours closure per day for maintenance, would provide a daily capacity of 60 trains in each direction.

It was a low-cost solution, with the merit of avoiding the principal controversial elements of the previous scheme: no Cheriton Terminal for road vehicle shuttle services; no high-speed rail link; more modest London terminal facilities and little or no threat posed to the expansion of the sea-ferries to meet future traffic growth.

On the departure of Richard Marsh, Sir Peter Parker had become Chairman of the British Railways Board. A man of considerable charm and personality, he was positively effervescent in his enthusiasm for the two goals he had set himself on taking over. One was the electrification of BR and the other, the Channel Tunnel. He foresaw the latter as an entirely State-funded enterprise, linking as it would the two State-owned railway networks. Subject as it was to the criticism of the everyday travelling public, BR would find it difficult to play a leading role in the construction of a Tunnel. Sir Peter was single-minded, however, and he nurtured the idea that the Government would eventually grant BR the ability to raise funds on the money market – a privilege already enjoyed by French Railways.

Bob Barron, a main board director, was given overall responsibility for Tunnel affairs; David Williams, who had earlier assisted Michael Bonavia, would be in charge of the day-to-day promotion of BR's new venture; and it was Richard Cottrell, MEP for Bristol, who christened the new tunnel proposal the 'Mousehole'.

British and French Railways found mutual compatibility within the scheme, but two relatively minor problems led to protracted discussions which were never satisfactorily resolved, as they were overtaken by time and events.

It had been mutually agreed that the single running Tunnel should be built to UIC gauge, with 25 k.v. overhead power transmission. BR had opted for the joint development of a new dual-voltage locomotive which would use the existing third rail dc power transmission on the London-to-Folkestone track, switching to overhead transmission on entering the Tunnel. However, for some reason which was never fully explained, SNCF objected to dual-voltage locomotives.

The other bone of contention was lavatories. BR opted for chemical toilets or, failing this, lavatories would be closed to passengers whilst travelling through the Tunnel. Although they were in use on some French trains, SNCF did not much care for chemical toilets. Accepting that a 35-minute journey through the Tunnel without access to lavatories was hardly a severe privation, SNCF was against the idea of locking and unlocking lavatories and

phlegmatically clung to the established practice of railways, even through tunnels, to evacuate the toilets between the tracks whilst on the move. BR, unsure whether SNCF was serious, or just bored with the constantly recurring subject, reminded it rather starchily that the Tunnel was 50 kilometres long and if this practice were upheld, it would soon become the largest sewer in the world – leaving aside the problems the daily maintenance crews would encounter! At the next meeting SNCF was all smiles – it had the solution! As the Tunnel was to be lit with strip lighting throughout, ultra-violet lighting tubes should be used, as the waste from the toilets was biodegradable!

In February 1979, Sir Peter Parker sent BR's preliminary technical and economic report to the Minister for Transport, who was now Sir William Rogers, following the premature death of Anthony Crosland in 1977. But the Labour Government's tenure of office came to an end with the General Election of May 1979.

The Conservatives returned to power with Margaret Thatcher as Britain's first woman Prime

Minister. It was soon made clear that public sector borrowing was regarded as a major contributing factor to inflation. The Government had inherited runaway inflation and a million and a half unemployed – both on the increase – and public sector spending would be curbed. Sir Peter, wisely recognising the constraints of the new monetary philosophy, did an overnight *volte-face* and announced that the British half of BR's tunnel would be funded by private venture capital and the French half through SNCF's investment budget.

Over the years supporters of the project had grown resigned to delays caused by General Elections and Ministerial reshuffles, particularly when they affected Transport. No new Minister in this Department could be expected to rate the Tunnel very highly in his order of priorities. This time would have been no exception – save that chance intervened and was given a nudge in the right direction.

Author addressing a meeting about the Channel Tunnel.

John Wells (later knighted), Conservative Member for Maidstone in Kent, provided the chance and also the nudge. He was Chairman of the Parliamentary All-Party Channel Tunnel Group, which had so ably acted as a source of information in both Houses of Parliament during the 1970s. He was holding a constituency meeting on the subject of 'Transport in Kent' and Norman Fowler, the new Minister of Transport[62], had agreed to attend. John Wells, recognising that the latest information on the Channel Tunnel would make a topical contribution, invited BR's David Williams and one of his associates[63] to speak. After the meeting, it so happened that they accompanied the Minister on the return train journey to London. The opportunity was too good to be missed and by the time they arrived Norman Fowler had been well briefed on BR's 'Mousehole'. Although not without knowledge of the subject he had not been fully conversant with the new low-cost and less controversial concept. He expressed great interest, but emphasised that the venture would have to be totally privately-funded. There would be no question of any Government participation by way of public investment, loans or financial guarantees.

Shortly afterwards, Norman Fowler announced to the House that he would initiate an immediate evaluation of British Rail's proposals and that a conclusion would be reached as soon as possible. Referring to the previous scheme, he said:

> The low-cost option – which is what the British Railways Board is putting forward – meets many of the objections that people have to the former scheme.

---

[62] Designated as Secretary of State in January 1981.
[63] The author.

In March 1980 British Rail produced a brochure entitled *Cross Channel Rail Link*, which initially had a selective distribution. In the foreword, Sir Peter Parker wrote:

> Behind this modest proposal by British and French Railways for a single-track rail link, there are great objectives. The railways can play a decisive role in the future economic development of Europe, as a highly energy-efficient transport mode. But the gap that separates the national rail networks of Britain and France must first be closed.

After BR had announced its plans, it was Norman Fowler's further statement in Parliament on 19th March, 1980, that unleashed the pent-up enthusiasms of a number of promoters who leapt at the opportunity to play a role in permanently linking Britain with France. Norman Fowler said that he had been examining preliminary proposals by British and French Railways and he awaited, with interest, the full proposal. Public funds would not be available, and he concluded:

> I look forward to receiving any specific proposals, including those on which British Railways are working which would attract genuine risk capital.

This invitation prompted a rash of competing ideas and schemes for a fixed link. At one time there were 13 different proposals for tunnels, bridges and combinations of both. All were variations on old schemes that, apart from refinements in design and construction, had first seen the light of day during the nineteenth century. At this stage, however, the only scheme which had been studied in any depth was the 'Mousehole'.

If the Tunnel had not once more been thrust into public prominence it is doubtful whether any of these alternative options would have attracted media attention. A hastily conceived outline plan enhanced by an eye-catching artist's impression served to keep the promoter's name before the public at little cost. For whenever the Tunnel was mentioned, other promoters would appear as contenders. The Government therefore appointed of a Select Committee for Transport[64] on 25th March, 1980, to refine the many proposals that had been submitted.

The Committee's task was to narrow the range of options on the capacity and form of a fixed link and to submit its recommendations to the House to enable Members to take an informed view. Norman Fowler also recalled Sir Alec Cairncross to

[64] Labour: Chairman Tom Bradley, Gordon Bagier, Sidney Bidwell, Neil Carmichael, Harry Cowans. Conservative: Stephen Dorrell, Den Dover, Peter Fry, Barry Porter, David Price, Gary Waller.

advise him as to the economic aspects of this abundance of choice.

So the courtship rituals began once more: the merry-go-round of committees, reports, new studies and re-studies of old studies – all this only to establish the *bona fides* of a serious suitor!

The sea-ferry operators rose to the challenge by replaying their theme song 'Anything you can do, we can do better and cheaper'. Ephemeral promises of cheaper, faster, bigger and more efficient ferry services, capable of coping with all future increased traffic demands, were nailed to the mast-head. The same promises were made in the 1970s, but once the threat of the Tunnel receded so did the promises. In the intervening years their tariffs had consistently remained higher than the inflationary rise in the cost-of-living index, with the profit from sales of duty free goods making a major contribution to their balance sheets. Even at this tariff level, as Sir Alec Cairncross said in his report:

> The ferry operators had no cause to regard 1980 as a particularly profitable year. On the contrary, the Chairman of European Ferries, Mr. Keith Wickenden MP, said in his Annual Report for 1980 that it had not been possible to raise charges enough to cover increased capital charges and other costs: ferry prices, he suggested, were 'dangerously low'. Similarly, Mr. Ian Churcher of P & O Ferries said that passenger and freight rates were both 20 to 25 per cent too low for comfort.

In competition: cross-Channel ferries.

Apart from this familiar grumbling undertow, there were new schemes for the Minister to note and for the public to ponder. George Wimpey Ltd. rested its case on a newly-formed partnership with Royal Volker Steven, a Dutch construction company. Its case was for an immersed tube tunnel, a construction technique in which its Dutch partner was a leading specialist. Wimpey admitted, however, to uncertainties of time and cost which would in all likelihood deter a backer. Tarmac Ltd. went straight back to 1975 – to the scheme begun, then abandoned, but which had been based on the most thorough prior investigation. Yet economic circumstances would suggest a more prudent approach: adopt three construction stages – let the Tunnel grow to meet increased demand – thus cushioning its impact on ports and ferries – and let first-stage revenue subsidise subsequent stages of construction. Then there were bridge schemes. Freeman Fox, consulting engineers, promoted 'Linkintoeurope' – a conventional suspension bridge designed by Bill Brown, one of the partners. And Pell, Frischman and Partners promoted the unconventional Eurobridge, which was a development of Frischman's 1960's proposal: instead of a suspended road-deck, it offered a suspended tube-tunnel containing four decks of motorway dimensions for road traffic. Both suspension bridge schemes envisaged a separate undersea-bed

tunnel for rail traffic. Technical considerations disabused any ideas of combined road-rail bridges.

Last time it had been French, now it was British: Redpath, Dorman Long, a wholly-owned subsidiary of British Steel, proposed a combination of viaducts, artificial islands and immersed tube tunnels. The stark problems remained, however: any bridgeworks attract huge maintenance costs, and the piers would be a permanent hazard to shipping.

In order to avoid the obstruction of mid-Channel shipping lanes, designs which transgressed the frontiers of proven technology were produced, such as unsupported central bridge spans of five kilometres[65] and hybrid combinations of bridges and islands with immersed tube tunnels under the main shipping lanes – all very high cost solutions.

The fact remained that railway-operated tunnels, for both road vehicles and classic passenger and freight rail traffic, retained an unassailable competitive edge over all other more extravagant concepts, which cost at least three times as much. Tunnelling is a tried and proven method of construction, well within the scope of known technology and could attract private venture capital in relation to quantifiable risk. Even without rail connections at either end, a rail-operated shuttle service for road vehicles remains the cheapest and quickest way of moving road traffic. As it would be connecting two rail networks, the additional through-put of classic rail passenger and freight could be achieved at little extra cost.

The Select Committee investigations continued into 1981. It was becoming evident that BR's 'Mousehole' scheme for a single six-metre-diameter tunnel dedicated to conventional rolling-stock would be condemned to total dependence on traffic generated by British and French Railways and would remain at the mercy of the capricious vagaries of future Government transport policies.

To lock what would become the most important transport route in the western hemisphere into such an inflexible and uncertain future was unthinkable. The conclusions of the Select Committee's final report[66], which was published on 6th March, 1981, reflected a shift of opinion away from the 'Mousehole', and opted for compromise.

It was recognised that BR was the only contender which had effectively done its homework, but the Committee abandoned the idea of a six-metre-diameter tunnel, and recommended that legislation be placed before Parliament for the initial construction of a single main running tunnel 6.85 metres in diameter, plus a service

---

[65] The longest unsupported span in the world is the Humber Bridge, 1.4 kilometres in length.
[66] Second Report from the Transport Committee Session 1980-81 *The Channel Link*, HMSO Vols I, II, III.

tunnel. It was conscious of 'the past trend towards road transport and the present majority preference for the road mode', and that it would be irresponsible to support a solution which precluded provision for road transport. Tunnel facilities would subsequently expand to accommodate a road vehicle ferry service. Rail only was abandoned.

The Report went on to consider the French Government's position. It had become evident to the Committee that the current attitude

> . . . was one of extreme caution with a marked reluctance to make any public commitment at this stage, in view of what the French authorities understandably regard as a previous breach of faith by the British Government in the cancellation of the 1974 tunnel project.

Norman Fowler confirmed that the French felt that it was for the British Government to take the initiative to demonstrate its commitment anew. He believed this to be perfectly reasonable: 'The ball is firmly in our court and historically I do not think anyone can blame them for taking that view.'

Two months later, the French went to the polls. On 10th May the Giscard d'Estaing Government was ousted and François Mitterrand was elected President as his Socialist Party swept to power with an absolute majority of 70%.

Giscard d'Estaing, with every justification, had personally remained ultra-cautious toward the revival of the project and nothing short of an irrevocable agreement with the British Government, with no escape clauses or fine print, would assuage his wounded *amour propre*. François Mitterrand, however, was to be much more receptive.

By this time, Tarmac's proposed three-stage construction of the abandoned 1974 project had been developed in greater detail and was submitted to Norman Fowler on 16th January, 1981. Eric Pountain[67] announced that discussions had been held with various organisations in France and elsewhere on the Continent, who had confirmed interest in supporting the scheme:

> *Stage One:* Completed by 1990, a single seven-metre-diameter main running tunnel and a 4.5-metre-diameter service tunnel for the passage of rail passengers

---

[67] Now Sir Eric Pountain, Chairman of Tarmac PLC.

and freight only. Cost £850 million.

Stage Two: Completed by 1996, the construction of road vehicle train shuttle service facilities operating from terminals adjacent to tunnel portals near Folkestone and Calais. Cost £450 million.

Stage Three: Completion by 2004 of a second seven-metre-diameter main running tunnel to facilitate two-way rail traffic. Cost £480 million. (All cost estimates at 1980 prices.)

Whereas Stage One – with the exception of the increase in the diameter of the main running tunnel to seven metres – was similar to the 'Mousehole' in operation and cost, it embodied a progressive flexibility for expansion with the future capability of transporting roll-on/roll-off road traffic, thereby serving a wider national interest. Although Tarmac's scheme would be entirely financed by private capital, with no recourse to Government finance or completion guarantees, it did – bearing in mind the fate of the last two attempts to build the Tunnel – require a Government indemnity against political cancellation.

In July 1981, Tarmac formed Channel Tunnel Developments 1981 Ltd. (CTD 81) and the following September was joined by George Wimpey in equal partnership[68]. Together with their respective Merchant Banks, Kleinwort Benson and Robert Fleming, they represented a powerful new grouping. The objective was no less than to sponsor, finance and manage the construction and subsequent operation of the Channel Tunnel.

Apart from general promotion, CTD 81 concentrated its attention on three broad target areas: Parliament, where the ultimate decision would be made; Kent, where the greatest opposition to the project could be expected; and France, where a fairly cool reception to any fresh dialogue on the Tunnel was anticipated.

South-East Kent had special problems. It had endured Channel Tunnel planning blight for nearly twenty years, with the Cheriton site at Folkestone – two-thirds of which was owned by the Department of Transport – having been earmarked for a future road vehicle shuttle terminal. There would be sensitive environmental issues and

Blots on the landscape of South-East Kent.

68 CTD 81: Chairman Tony de Boer (Chairman of British Road Federation), Alan Osborne (Director, Tarmac), and Tom Candlish (Chairman, Wimpey International) were responsible for the overall direction of the company. Tony Gueterbock (Wimpey) and Martin Hemingway (Tarmac) were responsible for day-to-day executive activities. The operational office of CTD 81 was at Wimpey's HQ, Hammersmith, London.

the Folkestone and Dover areas would have to bear the brunt of major disruption throughout the period of construction. Unemployment in the county was among the highest in the country, which understandably led to concern about the impact that any form of fixed link would make on the Port of Dover and the ferry operators, who were also the largest employers and biggest ratepayers in Kent.

The courtship rituals continued and further trials of strength were held to establish the worthiness and credibility of the suitors who were wooing the two Governments.

The venue of the annual Franco/British Summit meeting to discuss matters of mutual interest alternates between the two countries and on 10th and 11th September, 1981, it was held in London. It was François Mitterrand's first overseas visit as President, and his first meeting with Margaret Thatcher, despite the gulf in political doctrine that separated them, was very agreeable. It concluded with their agreement in principle that a fixed cross-Channel link demonstrated promise of considerable benefit to their respective national economies.

Any fears that the French 'would be playing hard to get' were removed at a stroke. Even after the debacle of 1975, France's Socialist President was not only prepared to listen, but to participate. There was no denying the President's enthusiasm, it made good political sense. The *Nord Pas-de-Calais* region was a Socialist stronghold and badly needed the employment that the project would provide. Furthermore, the overwhelming Socialist victory had introduced into public office many with no previous experience and the President wanted to demonstrate that his Government was just as capable as its Right Wing predecessor in undertaking enterprises of grand design.

No time was lost in putting the summit agreement into effect and the British and French Ministers of Transport announced on 21st September the formation of a Joint Study Group comprising officials of their Departments to study:

> . . . the type and scope of possible fixed links taking account of the interests of maritime transport with a view to advising both Governments on whether a scheme for a fixed cross-Channel link can be developed which would be acceptable to, and in the interests of, both countries.

Chaired on the British side by Andrew Lyall, Under-Secretary for International transport, and on the French side by Guy Bribant, *Chargé de Mission auprés du Ministre d'État,* the new Group met for the first time on 28th September, 1981.

Meanwhile, BR and SNCF, seemingly undeterred by the Select Committee's rejection of proposals for small bored tunnels,

Euroroute – the scheme submitted by Redpath, Dorman Long, a subsidiary of British Steel.

announced on 6th October that they were joining forces with the European Channel Tunnel Group (ECTG) to promote the six-metre-diameter 'Mousehole'. ECTG had itself submitted to the Select Committee *five* proposals for both bored and immersed tube tunnels, including one similar to that envisaged by BR and SNCF.

However, CTD 81's promotion of its project rapidly gained support in appropriate quarters of influence and with such third party endorsement soon took the lead position. The closest contender was Euroroute – the combination of viaducts, immersed tubes and artificial islands – although this was more a grandiose concept than a well-researched project.

Its new-found prominence was due to the appointment of Ian MacGregor as Chairman of British Steel in 1980. In an unprecedented move, Margaret Thatcher had engaged his services from the Merchant Bank *Lazard Frères* NY to put the State-owned house of British Steel in order. Lazards had played a financial role in the construction of the Chesapeake Bay Bridge Tunnel in the USA in 1964, which combined 15 miles of viaduct with some two miles of submerged tube tunnel in the middle to allow for the passage of shipping. MacGregor devoted considerable time to promoting the Euroroute scheme, comparing it with the Chesapeake Bay project – although anything less like the Chesapeake Bay Bridge Tunnel would have been difficult to imagine. Moreover, MacGregor took full advantage of his association with No. 10 Downing Street to overcome this advantage of happenstance, which left Euroroute's competitors having to work that much harder.

An announcement from Paris on 13th February, 1982, at first sight seemed to be at odds with the British Government's stated policy that any form of fixed link would have to be privately funded. President Mitterrand's election promises were to be redeemed. In the belief that Socialist doctrines, which had won the election, would prevail over economic reality, the new Government had embarked on a programme of nationalisation, which included the State ownership of private banks. This meant that any form of fixed cross-Channel link would have to be funded by a mix of French State funds and private sector financing on the part of the

UK. Such an *économie-mixe*[69] solution, was not insurmountable, just an added complication.

There was, however, nothing complicated with the result of the UK/French Study Group Report[70], published four months later. Having examined all the competing options for a fixed cross-Channel link, it concluded:

> The Group believes that the balance of advantage lies with twin bored rail tunnels with a vehicle shuttle constructed if necessary in phases. This solution would appear to be in the broad interests of both countries, since it would offer a secure means of transport, would be energy-saving in operation and would not adversely affect employment . . .

The Group rejected all other methods for a fixed link. Any project which offered 'drive-across' facilities would have the capacity to take *all* foreseeable traffic, even generated traffic. However, lorries carrying dangerous cargoes would not be allowed to travel through a road tunnel, and possibly would be forbidden access to a bridge. Any such scheme if adopted could only jeopardise the entire future of ferry operators.

In respect of bridges, immersed tube construction, and schemes which included artificial islands and ventilation shafts, the Group concluded that these were 'likely to provoke changes in the hydrology of the Channel . . . which could disperse existing sandbanks'.

As to the future of the sea-ferries, in the Group's judgement:

> The best solution from the point of view of both countries would be one which combined a fixed link of the kind proposed [tunnel] with a thriving maritime industry still carrying as much as, or more than, the traffic it now carries, in conditions of healthy and constructive competition.

The CTD 81 consortium received the Group's Report with some elation, by contrast to the depression which must have overwhelmed the competing promoters. Shortly after the Report had been presented to the two Governments for consideration, however, world events again took a hand.

Just as the British Cabinet was poised to discuss the Report, the news came that *HMS Sheffield* had been sunk in the Falklands conflict by the Argentine Air Force using a French Exocet missile. The Cabinet suddenly found itself in no mood to discuss cross-Channel links. This set-back could have been critical for the

---

[69] The *économie-mixe* solution never arose, as the French Government, faced with a failing economy, in 1981 reversed the State ownership of certain banks.

[70] Command 8561, UK/French Study Group Report, HMSO June 1982.

project's political momentum, had not some pretty fancy footwork in the British and French Departments of Transport salvaged the position.

Later in June the Ministers of Transport, now Charles Fiterman and David Howell respectively, issued a joint statement: they considered that the Study Group Report was essentially a technical appraisal; an examination of the schemes' financial implications was necessary before a decision was taken. On 11th August, 1982, they gave their blessing to a group comprising National Westminster and Midland banks, *Banque Nationale de Paris, Credit Lyonnais and Banque Indo-Suez*. This group was to examine all fixed link options, even those which earlier official reports, and latterly the Joint Study Group had found either technically unrealistic, or commercially unsound – being unlikely to attract private venture capital. This new study, which later proved to be flawed at the outset, resulted in Tunnel matters being shelved for nearly two years!

The UK General Election of 9th June, 1983, sustained Margaret Thatcher's Conservative Government with an overall majority of 144. Although inflation was gradually coming down, unemployment was approaching three million. The social and economic climate of over a decade earlier, when the project was approaching realisation, had vanished. It had been a period of full employment and no inflation. Political and public concerns were much more mundane: how much cement would be used in construction? Should scarce resources be diverted from higher national priorities such as the building of roads, schools and housing? Under no circumstances were Cross-Channel Contractors to tempt miners away from the Kent coalfields with 'riches beyond their dreams'. No one raised fears of a terrorist attack on the Tunnel – such acts were quite rare at the time. Equally, no one raised the fear of rabies or the question of safety, and the voice of the environmentalist was more muted. There had been concern about the rare Spider Orchid which grew on the Cheriton Terminal site, but it has since disappeared completely. Terrorism, safety, rabies, the environment, employment and the growth of tourism, were all matters which CTD 81 had to address in order to maintain its high public profile.

Model of Folkestone Terminal, 1983-84.

During February 1984, CTD 81, now based in Triangle House, Hammersmith, joined forces with two other groups who were promoting similar bored tunnel projects: the European Channel Tunnel Group (ECTG) led by Costain Civil Engineering, and the Anglo Channel Tunnel Group (ACTG), comprising Balfour Beatty and Taylor Woodrow Construction, who had been Cross-Channel Contractors in the 1970s project; these joined the new consortium under its title: the Channel Tunnel Group (CTG). This move greatly strengthened the bored tunnel lobby, bringing together the five leading UK construction companies and their banking associates, ready to collaborate with whomsoever would be representing French national interests.

Lord Shackleton, noted for his earlier involvement in the second attempt to construct the Tunnel, both as Leader of the House of Lords and Chairman of RTZ-DE, joined CTG on 4th April. Apart from his tunnel lore, he had been singled out by the Prime Minister to be one of her advisers throughout the Falklands conflict.

The Report of the five British and French Banks was at last completed and published under the title *Franco/British Channel Link Financing Group* on 22nd May, 1984. It concluded that the proposal of the Channel Tunnel Group: '. . . is the only scheme that is both technically acceptable and financially viable.'

CTG could not help but feel satisfied that once again its project had emerged head and shoulders above the competition. Yet the reception of the Bankers' Report was decidedly mixed, for the same reason that the UK Department of Transport gave it a rather jaundiced look. The Report's financial assessment concluded that even the CTG scheme – which had been thoroughly researched, both technically and financially, and the least costly of the alternative options – had little chance of being privately funded without Government loans of last recourse! A conclusion which would put an end to any form of fixed cross-Channel link. However, closer examination showed that the Banks had compounded an earlier measure of financial prudence which totally distorted the Report's conclusions. The Anglo/French Study Group (AFSG) Report of 1982 had accepted the six-year construction period estimated by each Tunnel promoter. It had, in its comparative analysis of capital costs and completion dates, assumed a 10% increase in cost and one year's completion delay. The Bankers' Report prudently added yet another year's delay, resulting in their base case being calculated on an eight-year construction period – with an overrun case based on a ten-year construction time. The effect of inflation and interest charges on loan capital over such time scales was astronomical and understandably the Report concluded that the project would need some sort of safety net or inducement by way of government loans of last recourse if it were to be privately funded.

Not unexpectedly, this shock conclusion resulted in a temporary hold on the project. However, CTG undertook an urgent re-examination of its financial formula and developed an accelerated construction programme based on minimum time, as opposed to a minimum cost basis. This would result in a 10% increase in the capital cost, but the project could be realistically completed in 4½ years. This new study reinforced the credibility of the original estimate of a six-year construction period. It also re-established the out-turn cost of the project to a level far below the assessments and predictions of the Bankers' Report. The shorter the period of non-revenue earning construction time, the more viable the project became and the more attractive to private investment.

Agreement on the new financial formula was finally reached with the Banks, and conveniently coincided with the State visit to Britain of President Mitterrand, 23rd to 26th October, 1984. CTG issued a statement. On the day of his arrival the President was greeted by the media with the news of CTG's new initiative: 'We can build the Channel Tunnel without Government help.' Sensing the mood, the President lost no opportunity to further the interests of the project in which he so firmly believed.

During the State Banquet on the night of 23rd October, the speeches made by her Majesty and the President were fulsome in their expression of mutual accord and goodwill, and in his reply to Her Majesty, the President said:

The author with Prime Minister Margaret Thatcher at the Conservative Party Conference, 1984.

Can our ties be said to be already sufficiently established and fruitful in the most varied spheres? probably not, and we cannot be content with them. And yet we have many joint tasks before us. May I mention – is it allowed? – the long-awaited Channel Tunnel, now a less daunting undertaking thanks to new technology . . .

During the remainder of his visit the President included references to the Tunnel in every official speech.

This new initiative by CTG – the brainchild of Tony Gueterbock – succeeded in bringing the private financing of the Tunnel back to the realms of the possible.

On 17th November Eric Parker, Chief Executive of Trafalgar House, the shipping and construction group which owned the Ritz

Hotel and the Cunard Shipping Company, announced that Sir Nigel Broackes, Chairman of Trafalgar, would be taking over as head of the Euroroute consortium from Ian MacGregor, who had recently become Chairman of the Coal Board.

The month of November 1984 was of great significance to the fortunes of the Tunnel. The annual Franco/British economic summit was held at Avignon on 30th November. At the conclusion of the conference, President Mitterrand and Prime Minister Margaret Thatcher issued a joint statement agreeing that: '. . . a fixed cross-Channel link would be in the mutual interests of both countries.'

The decision was immediately to be implemented. Yet another Anglo/French Working Party of Government officials would be formed, whose task would be '. . . to compile the specifications to be proposed to the companies interested in a cross-Channel link' within a deadline of three months.

At long last the two nations were confirmed in agreement. The Working Party – chaired jointly by Andrew Lyall, Under-Secretary of the British Department of Transport, and Paul Rudeau of the French Department of Transport – was to establish a framework for procedures and conditions governing the launch, funding, operation and safety of the project. Yet the all-important decision as to what kind of link had yet to be made.

Sir Nicholas Henderson, Chairman of CTG.

THE NEW BEGINNING

The project was once more poised on the threshold of a new initiative and the Channel Tunnel Group marshalled its resources in anticipation of the critical months which lay ahead. The formidable hurdles, both political and financial, which had beset earlier attempts would have to be cleared before tunnelling could start. A high level of political acumen would be required on both sides of the Channel. On 11th February, 1985, Sir Nicholas Henderson[71] was appointed Chairman of the Group.

The Channel Tunnel Group was a joint venture and its Board[72] comprised an impressive array of chairmen and directors from major UK companies, with an Executive Committee[73] responsible for the day-to-day conduct of its affairs. The National Westminster Bank was already a full member of the Group as Banker and Adviser, and Morgan Grenfell was appointed the Group's Merchant Bank Adviser. The project's traffic and revenue studies would be the key for a

---

[71] Sir Nicholas Henderson, having retired from a distinguished diplomatic career in 1979 as Ambassador to France, had been recalled by the Prime Minister to Washington, where he played a key role in explaining Britain's position to the American public and Congress during the Falklands War until 1982.

[72] See Appendix IV.

[73] See Appendix IV.

favourable response from private investment sources and Alastair Dick joined in March to formulate this all-important aspect of CTG's credibility[74]. Studies were commissioned from engineering consultants[75] and the congenial and very able Bill Shakespeare, with his solid Fleet Street experience with the *Guardian* and some twenty years as Northern Editor of *The Times,* was seconded from Tarmac to augment the press and public relations team. Michael Gordon, Deputy Chairman of Taylor Woodrow Management & Engineering, was appointed CTG's Managing Director on 1st April, 1985.

By contrast there had been no cohesive grouping of banking and construction interests in France. However, CTD 81 had opened discussions with several French banks and construction companies, of the calibre which would make them likely candidates for inclusion in the French Government's designated group, and these open-ended liaisons were now continued under the aegis of CTG.

To match the increasing tempo of events, rapid progress had been made by the Anglo/French Working Party. Since their appointments in early December 1984, members had completed an extremely detailed specification for a fixed cross-Channel link entitled 'Invitation to Promoters' which was published on 2nd April, 1985. Paul Quiles, the French Minister for Housing, Town Planning and Transport, gave it a 'send off' by declaring that the 'cross-Channel fixed link is no longer a sea snake', a somewhat odd comparison, but there was no doubting the sentiment. Nicholas Ridley, now Secretary of State for Transport, in a more formal statement in the House of Commons, reiterated the willingness of the Governments to take whatever steps might be necessary to facilitate the construction of a fixed link – that the project should be financed without support of public funds or Government guarantees – and reaffirmed that 'essential political guarantees would be provided' for the project.

This affirmation was of particular significance. The project had foundered, twice, through lack of political will and if a political guarantee was not forthcoming it would remain vulnerable. A Government's political guarantee by itself, with the best of intent, is not binding on future governments. It was therefore proposed that an appropriate form of words to overcome this would be included in the Treaty.

Copies of the 'Invitation to Promoters' were received by the four

---

[74] Following his appointment Alastair Dick and Associates, in conjunction with Wilbur Smith and the French company *SETEC-Économie,* entered into negotiations for the appointment of traffic consultants.
[75] Mott, Hay & Anderson, William Halcrow & Partners, Building Design Partnership and Ewbank Preace.

principal contenders – the
Channel Tunnel Group,
Euroroute, Eurobridge, and
Linkintoeurope – on the day
of publication. Each was to
submit its response to the
two Governments by 31st
October, 1985. Regrettably,
the competitive process was
yet again to be hampered

by the inclusion of fixed link schemes which had been dismissed
on technical and financial grounds by earlier independent studies.

At this juncture CTG's proposal for the project comprised two
main bored railway-operated tunnels, 49.4 kilometres in length (31
miles), 7.3 metres (24.6 feet) in diameter, bored at an average depth
of 40 metres (131 feet) beneath the sea-bed throughout and joined
to a central service tunnel, 4.5 metres (14.8 feet) in diameter, by
cross-passages every 375 metres (410 yards). There were to be two
crossover tunnels joining the two main running tunnels – situated
approximately one-third of the distance along the route from each
portal – to permit single tunnel operation during maintenance and
emergencies. Shuttle train terminals for the transport of road
vehicles would be sited adjacent to the portals at Cheriton, near
Folkestone, and Coquelles, near Calais. Phased construction was no
longer a consideration.

Despite the exhaustive technical and financial studies undertaken by
CTG, six months was little enough time to produce the wealth of
detail required by the Government's guidelines. It was an even
more formidable task for the promoters of the competing schemes
who had sustained their public profile with a project which had
been less thoroughly researched.

The move by the two Governments, demonstrating that a fixed
cross-Channel link was the subject of their serious consideration,
prompted a quick reaction from the Dover Harbour Board. On 29th
April it announced the formation of Flexilink – a consortium of port
interests and ferry operators from both sides of the Channel – for
the purpose of promoting and defending the role of the ferries on
the short cross-Channel routes. As before, ferries continued to
maintain that they could cope with all future cross-Channel traffic
growth and that, with higher utilisation levels, the new 'jumbo'
ferries and other economies, they could reduce tariffs.

Since 1980, new high-capacity ferries had been introduced, but
the shortest cross-Channel route was still the most expensive ferry
journey per mile in the world. Mile for mile it was cheaper to fly
Concorde to the USA!

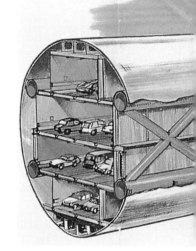

*Channel Tunnel Group
(above) and Eurobridge
(right). The two pictures
show projects that
competed, along with
Euroroute, for the
Mandate, 1985-86.*

The Anglo/French Study Group Report (1982) carefully considered the position of the sea-ferries and considered it unlikely that 'ferry costs would fall to the point which would weaken the economic case for most of the alternative fixed link options.' It also concluded that any fixed link providing a 'drive-across' facility would gravely jeopardise the future of the sea-ferries as it would 'divert traffic from shipping services to a point at which the latter would have difficulty in continuing operations on a financially viable basis.' It was the Group's judgement that railway operated tunnels of the type proposed by CTG would be the best solution for both countries, 'with the maritime industry still carrying as much, or more than the traffic it now carries, in conditions of healthy and constructive competition.'

Jonathan Sloggett, General Manager of the Dover Harbour Board, had, throughout the earlier years, competently presented the case for the ferries. In an interview with *The Financial Times* he said:

> The Board and other shipping interests do not oppose the building of a bridge or a tunnel. However the group believed a bridge, or tunnel should be obliged to compete fairly for freight and passenger traffic now carried on the ferries. Give us a fixed link with no Government guarantees and no hidden subsidies and with fair competition and we will beat it.

This was a fair and reasonable attitude and CTG believed that the ferry operators would continue to function in a competitive mode, alongside the Tunnel, conferring on the user – for the first time – freedom of choice. All forecasts of cross-Channel traffic growth concluded that it would double by the turn of the century and CTG studies revealed that the ferries' potential growth would plateau out when the Tunnel came into operation. The ferries on the short sea routes would continue their services, claiming a smaller share of a much larger market. Certain traffic, such as hazardous goods, would always be carried by ferry.

With the formation of Flexilink, however, Jonathan Sloggett went into the attack: the ferries would conduct a price war, cutting their tariffs to a level which would bankrupt the Tunnel. Nevertheless, they harboured the fear that the Tunnel would then be bought for a knock-down price which, in turn, would enable the new owner to wipe out the ferries by 'unfair competition'!

This 'as ye sow, so shall ye reap' approach was blind to who might end up the loser. For the victor immediately to cry 'foul', however, is merely obtuse, as devoid of reason as it is of understanding.

CTG had declared that it would be competitive with the ferries. Tunnel tariffs would be some 10% to 15% cheaper; there would be no advance booking; and the user would be carried further and faster by an all-weather day and night service. In the 1970s, when the ferry operators made the same price war threat, the British Channel Tunnel Company studied the possibility and concluded that the project was financially so robust that it could reduce fares by over 40% and remain viable.

Following the joint Governments' 'Invitation to Promoters', CTG, during May, entered into discussions with a number of French companies which had merged to form the company France-Manche. In format it was a mirror-image of CTG, comprising five leading French construction companies[76] and the three banks[77]. Two months later, on 1st July, 1985, Sir Nicholas Henderson on behalf of The Channel Tunnel Group and Jean-Paul Parayre on behalf of France-Manche signed a Co-operation Agreement. On the following day, the British and French Departments of Transport announced

Signing the Cooperation Agreement between CTG 81 and France-Manche, July 1985. Seated left: Jean-Paul Parayre; seated right: Sir Nicholas Henderson; standing left: Philippe Montagner, General Manager of France-Manche; standing right: Michael J. Gordon, Managing Director of the Channel Tunnel Group.

that a Joint Working Party of consultants would be appointed whose task would be to assess the submissions and act as advisers to the two Governments.

Following the merger of CTG/FM, the structure of the new grouping rapidly took shape. The five UK construction companies withdrew from CTG on 5th July to form Translink. Similarly, on 16th July the five French construction companies withdrew from France-Manche, to form Transmanche. These two consortia would be responsible for the building of the Tunnel at an initial capital cost of £2.33 billion (FF27.3 billion) at 1985 prices. The out-turn cost – with allowances for inflation and interest on loan capital during construction, plus £1 billion (FF11.7 billion) for unforeseen contingencies – was £4.3 billion (FF50.3 billion). In addition CTG/FM would raise £1 billion of public equity in three tranches.

During this important re-shuffle, the ranks of CTG had been further strengthened by the Mobil Oil Company and Granada joining as associate members; and the Midland Bank, the last of the

---

[76] Bouygues SA, Dumez, Spie Batignolles SA, *Société Auxiliare d'Entreprises* SA, and *Générale d'Entreprises.*
[77] *Banque Nationale de Paris, Crédit Lyonnais* and *Banque Indo-Suez* – the three French banks which had collaborated in the Franco/British Bankers' Report.

Anglo/French banking consortium, joining as a full member. To respond to the joint Governments' 'Invitation to Promoters', CTG/FM had assembled a team of outstanding calibre in very short order.

All the senior members of the team were seconded from the multi-disciplined ranks of the construction companies and banks which made up CTG/FM. Despite the fact that as individuals they bore an instinctive loyalty to their parent companies, the measure of overall co-operation reflected great credit on each member of this highly specialised group which rapidly forged a close knit team of total unity and purpose. Following six months of round-the-clock, intensive endeavour, CTG/FM's 2,500-page-submission was completed on time and formally delivered by Sir Nicholas to the Secretary of State for Transport; a similar ceremony was repeated in Paris by Jean-Paul Parayre.

Euroroute and Eurobridge also met the deadline, but the Linkintoeurope bridge scheme had fallen out of the race several months earlier. However, to everyone's surprise, another contender had also completed the course. This unexpected entry, presented by James Sherwood, Chairman of British Ferries, was called 'The Channel Expressway' and was the work of Sealink UK, a wholly-owned subsidiary of British Ferries. That James Sherwood and his company were leading opponents of any form of fixed cross-Channel link made the surprise even greater, and caused much embarrassment to Flexilink.

The Channel Expressway proposal was ill-researched and misconceived. Any one of the official reports produced during the previous four years should have convinced Sherwood of its obvious shortcomings. It was a proposal for two road/rail tunnels, each 11 metres (36 feet) in diameter, with a single rail track running down the centre of each tunnel. Every hour, road traffic would be halted to allow the passage through the tunnel of one diesel freight or passenger train. Furthermore, a 'special train service vehicle' would run ahead of the train, controlling the speed of the locomotive via an electric circuit to ensure that it did not overrun any road traffic remaining in the tunnel. Irrespective of all else, one diesel train per hour in each direction showed little

The Channel Expressway scheme.

faith in the future of railways.

Sherwood claimed that his scheme was the cheapest of all methods. Yet as long ago as the 1981 Select Committee it had been shown that to add one metre to the diameter of the Tunnel would cost an extra £100 million. Expressway was now claiming to be able to build two 11-metre road tunnels, sundry crossover tunnels at 500 metre intervals, by-pass tunnels every 1.7 kilometres, and ventilation shafts – and all for a highly improbable £2.1 billion!

As described earlier, ventilation is the main problem common to all road tunnels. The longest Alpine road tunnel, the St. Gotthard[78], approaches the limit for effective ventilation without multiple vent shafts. It is also close to the limit of the average motorist's capability to drive through a tunnel without becoming disorientated. Psychologists have identified what they call 'Highway Hypnosis', which they describe as a 'rigid cognitive set developing and creating a subjective notion of unchanging visual surroundings'. It is a confused condition which is as inexplicable to the lay mind as the description of its symptoms; and it is fraught with danger, in that a driver can experience a very real physical sensation that convinces him that his vehicle is moving sideways. Sherwood further claimed he had solved the ventilation problem with only two inshore vent shafts coupled with electrostatic precipitators, which – while cleansing the air of solid exhaust particles – would do nothing to disperse or reduce carbon monoxide levels. Until a method is discovered for controlling exhaust emissions of carbon monoxide, 'drive through' tunnels 50 kilometres long will remain a problem which only the future can solve.

A further complication was one of geology. The Lower Chalk, highly suitable for tunnelling, is simply not thick enough to contain tunnels of Sherwood's dimensions. They would inevitably intrude into the Gault Clay. The Gault Clay is under great pressure from the weight of the Chalk above. A tunnelling machine cutting through the bottom of the Chalk and into the Gault Clay would release this pressure and the softer Clay would immediately expand; to counter this uneven pressure the linings would have to be reinforced at enhanced cost. Boring within the limits of the Lower Chalk also releases stress and the Chalk expands slightly, but provided that the strength of the Chalk is roughly consistent all around the bore it maintains its structural integrity. This was demonstrated by the 1882 Beaumont Tunnel which withstood the passage of nearly a century without being lined.

The unexpected submission of the Sherwood proposal had further surprises in store. While the Working Party continued its vetting of the various schemes, James Sherwood sought and

---

[78] 16 kilometres (10 miles).

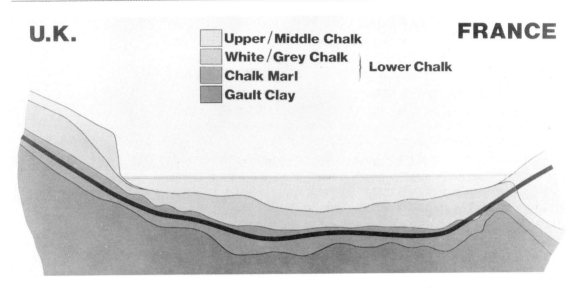

U.K.

Upper / Middle Chalk
White / Grey Chalk
Chalk Marl
Gault Clay

Lower Chalk

FRANCE

Schematic of Channel bed geology, 1983.

succeeded in making changes to his scheme on two separate occasions. Realising that he had underestimated the potential of BR and SNCF passenger and freight services and that his ventilation proposals were totally inadequate, he now proposed two 12-metre (39.4-foot) diameter road tunnels and two of smaller dimension for the railways. That these changes were accepted after the final submission date caused some concern amongst the other promoters, as the ground rules clearly specified by the Department of Transport seemed to have shifted. How did Sherwood get away with it? The Department played down the whole affair. One anodyne comment was: 'Well, he was something of a late-comer on the scene', which brought little solace to the other contenders who were playing according to the rules for high stakes. James Sherwood was certainly not without entrepreneurial talent; he had a flair for the theatrical and, as he had demonstrated, considerable influence in high places. The extent of his influence was not to be fully realised, however, until the extraordinary melée which was to take place at the eleventh hour.

The Parliamentary Transport Committee was once more asked to cast its eyes over the projects of the four principal contenders, and its report, published on 5th December, 1985, concluded that the CTG/FM scheme was the one which should be approved. It, too, was pretty scathing about Expressway which '. . . presented a greater challenge to credibility'.

William Hill, the turf accountants, was running a book on the outcome of the joint Government decision and on 10th December, 1985, the odds were as follows: CTG/FM was 1-2 favourite – Expressway was 7-4, Euroroute 3-1, and Eurobridge the 20-1 outsider.

Throughout 1984 and 1985 the three major promoters had carried the message of a fixed link up and down the country by way of presentations to industrial organisations, local authorities and the general public. As the competing groups frequently shared the same platform, a civilised and friendly rivalry had developed between them. The tall, commanding presence of John Low, the leading exponent of Eurobridge, and the genial Ken Groves, the practised advocate for Euroroute, became familiar figures. Audiences would listen and patiently watch slide shows and videos for two hours as the merit of each scheme was put forward. If the audiences were unsure and bemused at the start of these presentations, it was generally agreed that they were left completely confused by the end; but at least the principle of public consultation was honoured.

Author with Sir Nicholas Ridley, the Secretary of State for Transport, at Conservative Party Conference, 1984.

There was also consultation of another kind at the highest level. No doubt through the influence of Ian MacGregor, Ken Groves had given a number of presentations at No. 10. He never failed to let it be known that Mrs. Thatcher was not overly-fond of railways, and preferred 'a drive-across' scheme. Was Groves seeking to provoke and depress the opposition? The Downing Street Press Office denied that the Prime Minister had ever expressed such a view, and she subsequently denied it personally at a meeting with Sir Nicholas Henderson on 13th May, 1985. Sir Nicholas had expressed concern that sources in London and Paris were saying that she favoured Euroroute; she had then declared herself passionately in favour of a fixed link, but did not embrace any particular scheme, nor would it be for her alone to decide.

Behind the scenes, however, decisions and recommendations were being made which would remove, once and for all, any question of doubt. A Joint Working Party of British and French independent consultants had been appointed by the two Governments to assess and advise on the competing submissions. They met in Paris for the first time shortly before Christmas 1985, to find that they had independently arrived at precisely the same conclusion – the CTG/FM scheme was to be their joint recommendation.

But the rumour – if rumour it was – that the British Government favoured a 'drive-through' link persisted and exploded into prominence in early January 1986.

Sir Nicholas Henderson, in his book *Channels and Tunnels*[79], recounts vividly and graphically the events which followed a meeting between the British and French Ministers of Transport on 7th January, 1986. During discussions, Jean Auroux suggested that Sherwood's Expressway should be eliminated on the grounds that it lacked adequate French participation. To his utter surprise, the

---

[79] A collection of essays published by Weidenfeld & Nicolson, London.

suggestion was quickly challenged by Secretary of State Nicholas Ridley who unexpectedly went to some length to defend and extol the scheme's virtues. It so happened that Sir Nicholas and Jean-Paul Parayre were due to meet Ridley on the following afternoon. Prior to the meeting, Sir Nicholas – shrewdly sensing which way the wind was blowing – proposed that CTG/FM should keep all its options open, and give an undertaking that, if traffic growth warranted, and technology permitted, it would consider a drive-through project at some later stage. Parayre agreed and Sir Nicholas officially informed the Department of Transport of CTG's intent.

When members met Ridley, however, it seemed that the message had been garbled in transmission. The Secretary of State was under the impression that CTG/FM was proposing to join Expressway! Even after the misunderstanding had been spelled out to him, Ridley continued the meeting as if he had not heard.

Sir Nicholas's account of the next two critical weeks bears all the characteristics of having been scripted by Lewis Carroll as a real-life replay of *Alice in Wonderland*.

Week One
- Ridley says British Government would like a 'drive-through' scheme.
- Ridley, having met Sherwood of Expressway, seeks meeting with Sir Nigel Broackes of Euroroute.
- Sherwood releases a press story that Expressway has won the mandate.
- Prime Minister also under the misapprehension that CTG/FM is to join Expressway.
- CTG/FM issues statement to the effect that it is not joining Expressway.
- Press publish story of merger between Sherwood and Broackes.
- Bouygues SA, independently of CTG/FM, meets Sherwood and Broackes and proposes a three-way split: Euroroute would undertake the finance, CTG/FM would undertake the construction, and Sherwood would be the operator.
- Sir Nigel Broackes, at a meeting with Sir Nicholas, says he is not interested in joining Expressway – proposes that CTG/FM joins Euroroute.
- CTG/FM Board unanimous – its members are not prepared to join either Sherwood or Broackes.
- It is revealed that President Mitterrand favours Euroroute.

Week Two
- Sir Nicholas circulates Cabinet Ministers, setting out central features of CTG/FM's scheme, including readiness to build a 'drive-through' scheme in certain circumstances.
- Sir Nicholas meets Sherwood who says that Broackes is out

of the running and that Ridley is keen on CTG/FM getting together with Expressway.

- Result of Cabinet meeting – Government to stick to its original promise to come out in favour of one project, rather than organise an amalgam.
- Sir Geoffrey Howe, Foreign Secretary, cancels prearranged weekend at Chevening with Sir Nicholas, as it is embarrassingly close to the summit meeting at Lille at which the award of the mandate is to be announced.
- *The Times* publishes interview with Sir Nigel Broackes who is severely critical of Ridley for showing favouritism to Sherwood.
- French Minister of Transport furious with British Sunday Press for having reported ahead of time that CTG/FM had won the mandate.

Artist's impression of the French Tunnel Terminal which covers an area of 700 hectares – three times the area of the British Tunnel Terminal.

On 20th January, 1986, Prime Minister, Margaret Thatcher and President Mitterrand met at the Lille summit meeting, and in a joint statement it was announced that CTG/FM had been awarded the mandate for the development, construction and operation of a fixed link across the Channel. Both the President and the Prime Minister marked the event with expressions of goodwill and unity. The President spoke philosophically:

> France is delighted to bear witness to the fact that, when one has the will, it is always possible to unite people who are already drawn together by so many things . . .

The Prime Minister struck a more pragmatic note:

> We have made the right choice and passed a fundamental stage in the co-operation between the UK and France . . . the project is not the last word, it is the first step. It will be judged a thrilling undertaking . . .

The sudden rush of affection for James Sherwood's ill-conceived scheme was inexplicable and will probably remain a mystery. Its technical shortcomings must have been blatantly obvious to Nicholas Ridley as he was by profession a civil engineer. His insistence that the British Government wanted a 'drive-through' scheme totally ignored the opinions and unanimous conclusions of the teams of independent technical and financial experts which made up the Joint Governments' Working Party. He may have sought to appease those members of Parliament who still cherished fanciful thoughts of 'driving across'; but such wishful thinking could never have been realised by Expressway – certainly not by 1993, and probably not this side of the twenty-first century!

Justice was seen to be done with the publication of a British

Government White Paper[80] which was an assessment of the Anglo/French Working Party consultants who had vetted the competing projects. It revealed how the Governments had arrived at their choice of a Channel Tunnel. It was CTG/FM's impeccable credentials, both technical and financial, that had won the day. The reasons for the joint Government choice were summarised as follows:

1. it was the soundest financially;
2. it carried the fewest technical risks;
3. it was safest from the traveller's point of view;
4. it presented no maritime problem;
5. it was the least vulnerable to sabotage and terrorist action;
6. it had an environmental impact that could be contained.

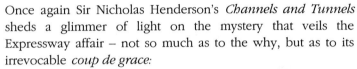

Once again Sir Nicholas Henderson's *Channels and Tunnels* sheds a glimmer of light on the mystery that veils the Expressway affair – not so much as to the why, but as to its irrevocable *coup de grace:*

I have learned subsequently that the French Government sent an emissary over to see Sir Robert Armstrong at a crucial point to make it clear that, although President Mitterrand favoured Euroroute, he would probably be prepared to set his preference aside and go for CTG/FM: but he would not withdraw his support for Euroroute so long as the British Government maintained its support for Expressway.

Euroroute had much more to offer the eye than a couple of holes in the ground; but such a choice would have been out of step with the President's concern for the welfare of the *Nord Pas-de-Calais* region. Over the decades, Calais had become resigned to the fact that eventually there would be a Channel Tunnel and it was believed that its membership of Flexilink was merely a tactical move to ensure that it could lay claim to regional 'distress' grants with the advent of a fixed link. However, the people of Calais would have manned the barricades if a 'drive-through' or 'drive-across' scheme had been chosen, as most of the traffic would be off and away onto the Autoroute without so much as a second glance at Calais. CTG/FM's Channel Tunnel, on the other hand, offered a pattern of traffic movement similar to that of road vehicles driving off the ferries. That was what they were used to, and what they wanted.

The Channel Link White Paper laid down the continuing arrangements for consultation and the next steps – the Treaty,

---

[80] Command 9735, *The Channel Tunnel Link,* HMSO.

Concession Agreement and Legislation. It also addressed the threat of cancellation which would bring comfort to both CTG/FM and investors. The Governments offered several assurances:

> . . . there would be certain guarantees once the Anglo-French Treaty and necessary national legislation had entered into force against political interference or cancellation. Also the promoter would enjoy full commercial freedom to determine his commercial policy . . .

This left a further period of perhaps a year and a half during which the project would be at risk, but the Governments' commitment had been made and the way ahead, once the instruments of procedure were in place, was clearly defined.

The eleventh hour gesture which CTG/FM had made towards a 'drive-through' scheme came home to roost, as it was included in the Concession Agreement, but not in such a way as to cause CTG/FM undue concern. The commitment was to undertake an examination into a 'drive-through' tunnel by the year 2000 when such a link might have become both technically and financially viable. If it proved viable, CTG/FM had the option, up to the year 2010, to put it into effect. If, however, it failed to do so, the Governments would be free to issue a general invitation for the construction and operation of a 'drive-through' link which would have to be financed without the support of public funds, or Government financial or commercial guarantees. Moreover, to avoid undermining the returns to investors in CTG/FM, the Governments would undertake not to facilitate any other form of fixed link before the year 2020.

On 12th February, 1986, President Mitterrand and Prime Minister Margaret Thatcher formally signed the Anglo/French Treaty in the ancient city of Canterbury, Kent, and this marked the beginning of Parliamentary procedures which had to be completed before the Treaty could be ratified.

The British Parliamentary legislative procedure would again be conducted through a Hybrid Bill[81] – a rather laborious and time-consuming process and one which, because of the tight timetable of Parliamentary affairs, was not overly popular in the House of Commons. It has, however, two important virtues. An ordinary Bill has to be passed within a Parliamentary year and, if it does not complete its procedural steps within this timetable, it goes by default. A Hybrid Bill overrides this time element and can be carried forward to the next Parliamentary year. Should there be a General Election during the Bill's passage and a Government of a different

---

[81] A Bill which involves both public and private interests.

Prime Minister Margaret Thatcher and President Mitterrand signing the 1986 Treaty in Canterbury.

political persuasion returns, there is a custom whereby it can be continued under the new Government. This was demonstrated when the Channel Tunnel Hybrid Bill was being processed through Parliament in 1974, when there were two General Elections. In February, the Labour Government, led by Harold Wilson, took office and adopted the Bill. In the following September, Wilson again went to the polls in the hope of being returned with an increased majority. He succeeded and again carried the Hybrid Bill forward.

The signatories of the 55-year Concession Agreement on 14th March, 1986, were Jean Auroux, *le Ministre de l'Urbanisme du Logement et des Transports*, representing the French State; Nicholas Ridley, Secretary of State for Transport in the Government of the United Kingdom; Jean-Paul Parayre, President of France-Manche; and Sir Nicholas Henderson, Chairman of the Channel Tunnel Group, on behalf of the two Tunnel companies.

There was a particularly starry-eyed reflection in the White Paper which carried more than a tinge of irony:

> Historically Britain's island status has often been an advantage.
> Today it is a practical and economic hindrance to closer links
> with Europe . . .

It had taken 153 years to finally exorcise the spectre of defence considerations, and a further 31 years to appreciate its passing. Irrespective of the doubtful credibility of defence objections, the practical and economic advantages of a fixed link with the Continent could have been no less important to the nation at any time throughout the project's history.

There was an even greater irony in that the project had come full circle. The present Tunnel scheme is similar in essentials to the proposal made by the Channel Tunnel Study Group in 1960 at an out-turn cost of £130 million. Fifteen years later, at the time of its abandonment in 1975, the cost had increased to an out-turn cost of £850 million. Ten years further on, the out-turn costs had escalated to £5 billion. Twenty-five years of political prevarication by a succession of British Governments had a costly consequence. Inflation will have played its role in these increased costs, but there could be no greater attraction than to pay off a 1960's debt in 1985 money terms.

Sir Nicholas Henderson resigned as Chairman of the Channel Tunnel Group on 31st March, 1986. During his eight-month term of office, he had successfully achieved the target he had set himself. One of the project's major achievements was the merging of five British and five French construction companies into a single, unified entity. This integration – with each company having different procedures, systems and ways of working, complicated further by national differences – was a prodigious accomplishment. Following the resignation of Sir Nicholas, Lord Pennock[82] became Chairman of CTG.

Opening of Information Centre, Tontine House, Folkestone, February 1986.

Two days prior to the signing of the Anglo/French Treaty, CTG opened its Information Centre at Tontine House in Folkestone – strategically located in the heartland of opposition to the Tunnel. As with the advent of the Hybrid Bill in Parliament and the absence of a Public Enquiry[83], Flexilink sought every opportunity to fuel mounting public militancy towards the Tunnel. The absence of a

---

[82] A former Deputy Chairman of ICI, with responsibility for Europe, Lord Pennock was currently President of the European Employers Confederation; and as Chairman of BICC had been closely involved with Balfour Beatty, one of the Tunnel construction companies.

[83] Under Public Enquiry procedures the Secretary of State for Transport is the final arbiter. He can either accept or reject the findings of an Enquiry – his decision being final and binding. Under the Hybrid Bill it is Parliament, the highest tribunal in the land, that makes the decision. Those affected have the opportunity to state their case before a Select Committee of MPs and again before a Select Committee of members of the House of Lords. These hearings, as well as the decision-making process, are all made in public.

Public Enquiry was ideal for this purpose, although the public was able to put its objections to Select Committees. Flexilink exploited the fact that it was not generally understood that petitions dealing with the same objection, word-for-word as some of them were, could not all be heard. It would be a time-wasting nonsense to subject the Select Committees to the repetitive exercise of hearing, say, 25 individual petitioners re-iterate the same thing on the same subject. In these instances the Chairman would ask them to appoint one of their number to represent their common concern. The same would happen in a Public Enquiry – but the bruised egos of those who were not chosen to represent their fellow petitioners remained bruised.

CTG's Information Centre, under the capable wing of Tony Gueterbock (later Lord Berkeley), played an invaluable role in allaying the fears of local communities. France-Manche, who had no organised opposition to contend with, opened its Information Centre in the rue Mollien in Calais some four months later.

On 6th April, CTG left Triangle House, Hammersmith, which had been the setting for the positive revival of the project, and moved to Portland House, Stag Place, London, where it was joined by France-Manche SA. Combining the British and French staffs of CTG and FM, Portland House became the Headquarters of both Companies, operating under the unifying title of *Eurotunnel.* On 2nd August and 18th November, 1985, Eurotunnel SA and Eurotunnel PLC had been incorporated respectively in anticipation of CTG/FM being awarded the Concession. Eurotunnel SA acquired France-Manche on 18th December, 1985, and Eurotunnel PLC had acquired CTG on 30th May, 1986. It was essential in every respect that the financing and construction of the project should be undertaken by a totally integrated Anglo/French Company with a Board of Directors in both Britain and France. It also ended the search for the missing piece of the mosaic. No longer was the Channel Tunnel a project looking for a client – henceforth the client would be Eurotunnel. On 18th October, 1985, the interests of Translink and Transmanche GIE had been integrated under the new unifying company title of Transmanche-Link (TML), which would be directly responsible to Eurotunnel for the construction of the project. It would also be the central purchasing agent for all the essential equipment and materials necessary to construct the Tunnel, from both sides of the Channel.

The Hybrid Bill had its First Reading in the House of Commons on 11th April, 1986, and its Second Reading on 5th June. During the intervening period Jonathan Aitken, Conservative MP for Thanet in Kent and champion of Flexilink, tried to get the Bill thrown out on

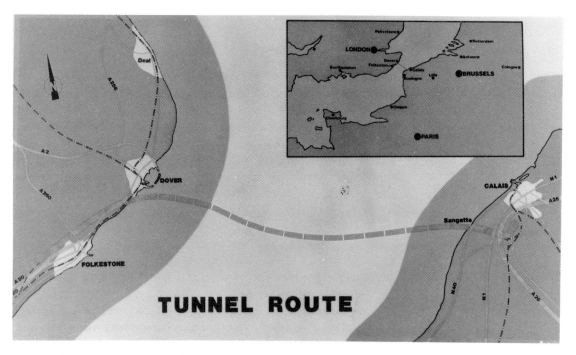

TUNNEL ROUTE

the technicality that it had contravened a certain procedural formality. He was overruled: the Second Reading was passed with an overwhelming majority – 309 for, 44 against – and the House of Commons Select Committee began its hearing of petitions. Not all petitions were necessarily opposed to the project; many simply sought safeguards to protect their interests during construction and operation.

The Select Committee received a record 4,866 petitions, which Flexilink claimed as evidence of the overwhelming opposition to the Tunnel. Flexilink omitted, however, to mention how some of the petitions had been obtained. There were several reports of supporters of Flexilink canvassing passers-by and handing them ready-made petitions to sign. This brought a public rebuke from Alexander Fletcher, MP for Edinburgh, Chairman of the Select Committee.

The Tunnel route from the Folkestone terminal at Cheriton to the Coquelles Terminal near Calais.

MONEY MATTERS

Eurotunnel had earmarked 10th July, 1986, for its launch of Equity 2. This issue sought to place privately some £200 million, against a tight deadline. The Concession had been signed a bare three months ago and financial advisers had underestimated the degree of information required by the international banks. The British Government's espousal of the privatisation of State industries had broken the tradition of share flotations being advertised solely in fine print in leading newspapers. Promotional hype and television advertising now ruled. Already British Telecom and British Gas had become the property of thousands of shareholders, many of them first-time investors.

Furthermore, Eurotunnel's shares had some unusual features: no dividend for at least seven years; shares had to be purchased in pairs, one British and one French – ownership of a share in half a Tunnel being pointless; and at £24.00 the pair they were expensive. Moreover, they were entirely different from shares in State industries which were going concerns, with very tangible assets; Eurotunnel's only assets were a piece of paper entitled 'Concession' and an as yet unratified Treaty, both of vital importance in their own right, but of no great intrinsic value. Eurotunnel therefore decided to postpone the launch date.

During this period some 14,000 visitors passed through the Information Centre at Folkestone, and Eurotunnel continued to play its part as a member of the Kent Joint Consultative Committee[84]. One problem for the Committee was 'What was to be done with the chalk spoil?' The UK share of the excavated chalk from the Tunnel would be of the order of 4.6 million cubic metres; and only one million cubic metres would be needed to level the 142-hectare (350-acre) site for the road vehicle shuttle service terminal at Cheriton. This would be deposited on the site via the land section of the Tunnel, which would be bored inland from Shakespeare Cliff – but how to use the remaining 3.6 million cubic metres?

Eurotunnel's original proposal was to extend the lower working site at Shakespeare Cliff seaward by depositing the chalk spoil behind a series of sheet piling containment areas, reclaiming a stretch of land some 1.5 kilometres (one mile) long, embracing an area of 26 hectares (62 acres). On completion, the working site would be landscaped to complement the natural area and provide an attractive amenity for Dover.

The Mitchell Committee was besieged with over 70 different suggestions for the disposal of the chalk spoil. For whatever purpose, it had to be close to a railhead as the environmental impact of transport by road was unacceptable, but in the end nothing matched Eurotunnel's reclamation proposal at Shakespeare Cliff.

Eurotunnel, on 15th May, 1986, signed the order for TML to start preliminary works, and the Construction Contract between Eurotunnel and TML was finally signed on 13th August. It stipulated that the Tunnel would be constructed on a target cost basis – the buildings and all related infrastructure of the terminals at Folkestone

---

[84] Also known as the 'Mitchell Committee'. It was set up by David Mitchell, Minister of State for the Department of Transport, and held its first meeting on 25th February, 1986.

and Coquelles, plus all the fixed equipment and mechanical and electrical elements of the tunnels, to be built on a Lump Sum basis. The locomotives and road vehicle shuttle rolling-stock were described as procurement items, which meant that TML, with Eurotunnel's agreement, would sub-contract their supply and installation. In essence, the contract was a turnkey project; all main construction would be completed by the Autumn of 1992, leaving some six months for TML to commission the system and hand it over to Eurotunnel in the Spring of 1993 as a going concern.

The original shareholders of CTG/FM formalised their Equity 1 investment of £44 million on 1st September, 1986. As a result of the

Preliminary investigations: the Submarine Continental Railway Company's No. 1 Heading was the first trial boring in 1880. Starting at Abbots Cliff, it was seven feet in diameter and covered a distance of 897 yards.

shortage of funds caused by the postponement of Equity 2, the British and French Banks in the consortium extended a further £9 million loan facility to Eurotunnel. On 25th September, the Joint Board of Eurotunnel PLC and Eurotunnel SA was announced, with Lord Pennock and André Bénard[85] designated as Joint Chairmen.

The Board[86] was an impressive grouping of leaders of industry and commerce, embracing both British and French interests. It included two tunnellers of long-standing: Sir Alistair Frame, now Chairman of Rio Tinto Zinc Corporation PLC, who had headed the Project Management team of the last attempt to build the Tunnel in the 1970s, and Sir Frank Gibb, Chairman and Chief Executive of Taylor Woodrow PLC which, at that time, had been the other half of Cross-Channel Contractors, the company which had started with such high hopes to construct that ill-fated endeavour.

Preliminary investigations: the underground stabling for pit ponies, Abbots Cliff, circa 1880.

[85] In addition to his well-known association with the petrochemical field, André Bénard was a Senior Adviser to *Lazard Frères*, a Board member of INSEAD, Director of *La Radiotechnique SA*, Chairman of the European Studies Unit and CEPS in Brussels and a director of Bekaert NV, Belgium.

[86] See Appendix IV.

The Equity 2 prospectus was issued on 21st October. The Units – each consisting of one share in Eurotunnel PLC and one in Eurotunnel SA – were not divisible, as it was intended that Eurotunnel would remain a private venture, and this proviso gave some protection against future nationalisation. The 8,583,334 Units would cost £12.00 sterling plus FF120 each, and should raise £103 million and FF1,030,000, the equivalent of £200 million.

Eurotunnel now joined in the new game of media hype by commissioning a 30-second video targeted at the Home Counties where the majority of fund managers lived, to be shown on Commercial Television at a cost, including prime time weekend screening, of over a million pounds. After two or three showings the Independent Broadcasting Authority (IBA) ordered its withdrawal. Jonathan Aitken, MP, had been busily devilling. He had lodged an objection with the IBA on the grounds that the screening of the video represented improper influence on the members of the Select Committee who were in the process of examining the Hybrid Bill petitions.

Jonathan Aitken's intervention had all been part of Flexilink's new campaign, directed specifically at the City and designed to undermine investor confidence in Eurotunnel shares. In reality it was more of an irritation than anything else. The video was a sad little production, completely unoriginal. Its sound-track comprised of little more than a voice-over declaring, with all the stridency of a crowd of football hooligans – 'Look out, Europe, here we come!' Apart from carrying the Eurotunnel logo, it bore no relation to the project whatsoever. Feelings within Eurotunnel were to say the least mixed, but the company was under pressure from its financial advisers and the banks to 'play the game' according to the new perceptions of launching company share issues – otherwise the money men could give no assurances as to the outcome of Equity 2.

The day after the issue of the prospectus, the City predicted a shortfall of some 10% in Equity 2 issue. Its cancellation in July, its high price, unusual format, long pay-back period, and the uncertainty in the market following the Government's sell-off of State industries, all contributed to a growing scepticism. Flexilink's muddying of the waters did not help the position.

Yet when dealings in the Equity 2 issues closed on 28th November, Eurotunnel had raised £206 million, and when the total was adjusted for rates of exchange, it reached £212 million. So much for the City's predicted shortfall.

Following the completion of Equity 2, Sir Nigel Broackes, Chairman of Trafalgar House, who had led the unsuccessful Euroroute consortium, accepted the invitation of the Co-Chairmen of Eurotunnel to join the Board.

Lord Pennock Co-Chairman of Eurotunnel.

The closing months of an eventful 1986 witnessed the start of preliminary works at Sangatte by TML on 30th October. On the same day TML commissioned its own manufacturing facility for the production of pre-cast concrete tunnel lining segments on a 70-acre site on the Isle of Grain, which lies between the River Thames and the Medway.

The turn of the year brought teething problems which the media were quick to exploit. Unfortunately this early fall from grace tended to dog Eurotunnel for the rest of the year, in marked contrast to the almost unanimous support that the project had enjoyed from the press for some 30 years previously.

The reason for the media raising a question mark against Eurotunnel's credibility was the resignations of the Chairman, Lord Pennock[87]; the Deputy Chief Executive, Michael Julien; and Board member Sir Nigel Broackes[88], all within a matter of days. There was no connection between these resignations; if they had happened at intervals none would have raised more than a passing ripple of interest. Alastair Morton[89] succeeded Lord Pennock as Co-Chairman of Eurotunnel on 20th February, 1987. The Board of Eurotunnel was strengthened with four additional appointments on that day: Bernard Auberger[90], Renaud de la Genière[91], Robert Lion[92], and finally Sir Kit McMahon[93].

Meanwhile, matters were making agreeable progress in the Houses of Parliament. The Hybrid Bill had completed its Standing Committee examination, had passed the Report Stage and had reached the Upper House where it had its Second Reading. The House of Lords Select Committee started to hear petitions on 2nd March.

---

[87] Lord Pennock, who became Co-Chairman in the previous year, had understood that his role would entail only a few days a week, but the mounting pressures of Eurotunnel had coincided with an increase in demands upon his time and responsibilities elsewhere and he therefore, with reluctance, decided to stand down. His statement was issued on 10th February.

[88] Sir Nigel Broackes resigned from the Joint Board on 2nd February. He would be missed – he had been a worthy adversary and with his talents had much to offer.

[89] Alaistair Morton had been the first Chairman of the British National Oil Corporation, and had held a number of directorships in the steel and engineering industries. Since 1982 he had been Chief Executive of Guinness Peat, the banking and financial group, and became Chairman of that company on the same day that he took up his position with Eurotunnel. Guinness Peat was later taken over by Equiticorp Holdings – a marauding group of financial buccaneers from New Zealand. Morton was touring the financial centres of the world at the time and, as a result, lost his chairmanship. He remained, however, with Eurotunnel.

[90] Chief Executive Officer of the *Caisse Nationale du Crédit Agricole.*

[91] Currently Chairman of the Board and Chief Executive Officer of the *Compagnie Financière de Suez* and of *Banque Indo-Suez.*

[92] Chief Executive Officer of the *Caisse des Dépôt et Consignations.*

[93] Chief Executive and Chairman designate of Midland Bank PLC and a former Deputy Governor of the Bank of England.

French Parliamentary affairs – whose procedures are far simpler and less time-consuming than the British – were also on the move. The *Déclaration d'Utilité Publique* had been completed on 1st August, 1986. This is a form of Public Enquiry which was a necessary step before the Anglo/French Treaty could be ratified by the Government.

Soon after taking office as Co-Chairman of Eurotunnel, Alastair Morton was confronted with a dilemma. Each passing day was fast narrowing Eurotunnel's room for manoeuvre – the problem being when to launch its third and final tranche of Equity. There was no knowing when the Hybrid Bill would complete its passage through Parliament. An extended sitting of the Select Committee in the House of Lords could delay events until the end of the year, since Parliament rises for the Summer Recess in July and does not resume its sessions until the end of October. There was also another General Election in the offing. The international consortium of some 50 banks,[94] whose agreement in principle to finance the project had been such a trump card in the award of the Mandate to CTG/FM, had since confirmed the commitment of £5,000 million of loan capital, but had stipulated that the loan could not be called in, or drawn down, until Eurotunnel had raised £1 billion of equity capital. Equity 1 and Equity 2 had jointly raised £250 million and Equity 3 would complete the total equity requirement. Eurotunnel, however, was restricted from any moves to effect completion until the Hybrid Bill had been passed and received Royal Assent, and the Anglo/French Treaty had been ratified.

Alastair Morton's solution to this ever-tightening knot of circumstance was to conclude the Usage Agreement with British and French Railways which would, in effect, guarantee 50% of the revenue of the Tunnel. On this basis the £5,000 million of loan capital could be called in, and a date for the launch of Equity 3 postponed to much later in the year.

Alastair Morton,
Co-Chairman of Eurotunnel
from 1987.

THE ROLE OF THE RAILWAYS

For the past three years, CTG had undertaken an annual tour of all the major cities throughout the country to maintain public awareness of the project. This tour, known as a 'roadshow', took the form of an exhibition combined with a day's seminar, to which local authorities and industrialists were invited to hear about the benefits of the Tunnel. Latterly the focus had been on the £1,500 million – £750 million on each side of the Channel – which TML would be spending on the essential materials and equipment needed to construct the Tunnel. In addition, British Rail would be

---

[94] See Appendix IV.

investing a further £400 million: £200 million on the up-grading of the existing Folkestone-to-London railway tracks, a new international station at Ashford in Kent, and a passenger terminal at Waterloo; and a further £200 million on a joint venture with SNCF to design and construct a new generation of high-speed trains dedicated to the Tunnel. There would, therefore, be over £1,000 million available to be spent in the UK on orders for materials and equipment.

Whereas it had never been claimed that the Tunnel would be the answer to unemployment, a billion pounds pumped into the economy over a relatively short period would make a significant contribution, as orders from TML would not only sustain existing employment, but create thousands of new jobs. As 80% of the UK's industrial and manufacturing strength lies 'to the north of Watford' – the hypothetical divide between the 'depressed North and the 'prosperous South' – companies in the Midlands and the North would be the most likely sources of supply.

During construction there would be upwards of 5,000 people employed at peak periods, principally in the Folkestone/Dover area, and during operation some 3,000 would be employed at the Folkestone Shuttle Terminal in operating and maintaining a no-advance booking, 24-hour, 365-day, all-weather service.

There was to be no new high-speed track between London and the Cheriton Terminal, which had caused such a rumpus during the closing stages of the 1975 scheme. BR, by ceasing to operate time-tabled boat trains and with other diversions, could muster 33 pathways daily.

The BR/SNCF Tunnel-dedicated, new generation of multi-voltage, high-speed locomotives and passenger rolling-stock would be capable of travelling on both UK and Continental rail networks. The new trains would achieve journey times of London to Paris: 3 hours 15 minutes; London to Brussels: 2 hours 55 minutes; which would be very competitive with the airlines, city centre to city centre. The new locomotives would travel from the Waterloo Passenger Terminal at 100 m.p.h., picking up power from the third rail 750 volt dc system, and transferring, on the move, to overhead transmission as they passed through the Folkestone Terminal into the Tunnel. Emerging from the French portal they would continue on overhead transmission at speeds of over 180 m.p.h. on the new *TGV-Nord* track via Lille to Paris or Brussels.

Progress is a much over-worked word. It is a little known fact that in 1924 it was possible to fly from London to Paris in a total journey time – city centre to city centre – some twenty minutes faster than is achieved today! This service operated between Croydon airport and Le Bourget. The speed of aircraft in those times – 'downhill and with a following wind' – was well below the take-off speed of most modern aircraft; and the drive to Croydon and from Le Bourget into Paris could scarcely have been a traffic problem. Still, travellers in the South-East destined for Paris by Tunnel would be making real progress. The journey time from Ashford International to the *Gard du Nord*, Paris, would take just over two hours – the fastest time ever achieved by a scheduled public service by land or by air.

The 1987 'roadshow' took the form of a jointly-sponsored BR/Eurotunnel exhibition train. It was given a 'send off' by John Moore, the new Secretary of State for Transport, at Paddington Station on 1st March, 1981, before departing on a 'whistle stop' tour around the country. John Moore also unveiled, on 20th March, full-scale models of Tunnel shuttle wagons which had been built by TML at the Old Railway Works at Ashford. These full-scale models were 'working tools' to enable TML engineers and designers to test loading techniques and make preliminary assessments of size and appearance before arriving at definitive design. They portrayed the three types of shuttle wagons which would transport all road vehicles between the shuttle terminals. When in operation they would be confined to the shuttle circuit as their sheer size – 5.6 metres (18.4 feet) from track to roof – would not permit their movement on BR or SNCF rail networks.

The dimensions and methods of loading and unloading shuttle trains would be based on the drive on/drive off principle. The configuration of all three types of single- and double-deck shuttle trains would be the same, e.g. a rake of 12 carrier wagons, plus a loading/unloading wagon and a locomotive at each end. At peak periods 'a train' would comprise two rakes.

Comparison of loading gauges.

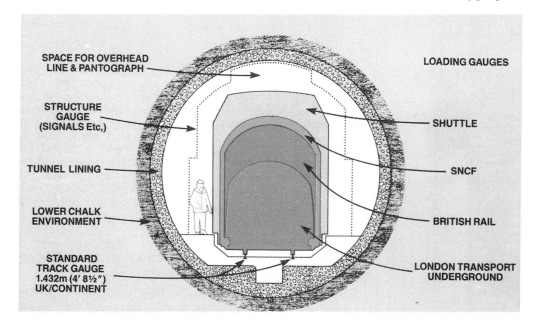

SPACE FOR OVERHEAD
LINE & PANTOGRAPH

STRUCTURE
GAUGE
(SIGNALS Etc,)

TUNNEL LINING

LOWER CHALK
ENVIRONMENT

STANDARD
TRACK GAUGE
1.432m (4' 8½")
UK/CONTINENT

LOADING GAUGES

SHUTTLE

SNCF

BRITISH RAIL

LONDON TRANSPORT
UNDERGROUND

The frequency of the shuttle services in the early years would be every ten minutes at peak periods of demand, every 20 minutes in off-peak, and every 30 minutes throughout the night. The design of the signalling and control system would be able to accommodate 30 trains per hour in each direction. No advance booking would be necessary as the system offered the equivalent of an on-demand service. The main road corridor out of London would be the M20, connecting directly to the Folkestone Terminal. The main roads to and from the terminal at Coquelles would be either Auto-route A26, or trunk roads such as the RN1 and the A26 from Paris.

The Coquelles Terminal, about one kilometre south of Calais, was to be three times the area of the Folkestone Terminal, which, physically constrained by the boundary of the Southern escarpment of the North Downs on one side and the M20 motorway on the other, severely limited the space available for essential facilities. It would initially have ten loading/unloading platforms, ultimately increasing to 16, to meet traffic demands of the future. Having passed through toll booths, drivers and passengers would be able to pause for refreshments or shopping, or pass straight through to British and French frontier controls, based on the 'free exit' principle, which reduces official procedures to one delay. Departing vehicles would pass first through UK customs and emigration and then immediately through French immigration and customs before boarding the shuttles. On arriving 30 minutes later at the Coquelles Terminal, drivers and passengers would be free to continue their journey without further delay. Procedures would be reversed on the return journey to the UK.

BR's passenger and freight services would join the terminal via a junction with the Folkestone-to-London main line to the east of Saltwood Tunnel. Alongside the rail link into the terminal would be a number of sidings at Dollond's Moor for outward-bound freight trains, providing facilities for customs inspection, changing of locomotives, and to hold trains awaiting a pathway through the shuttle service.

Situated at the western end of the Folkestone Terminal, on the fringes of the rail loop which encircles the area, are the villages of Newington and Peene. Although the top end of the loop would be underground in a cut-and-cover tunnel, and the villages screened off by trees and sound bunds, they would nevertheless be subjected to noise from the terminal 24 hours a day. There were a number of objectors to the Tunnel in the Folkestone area, but the residents of these two villages were in the front line of an environmental disturbance which would drastically devalue their property. The Channel Tunnel Group offered to purchase the properties of the residents of the two villages at the current market value should they wish to move. The offer would remain open for ten years and if the

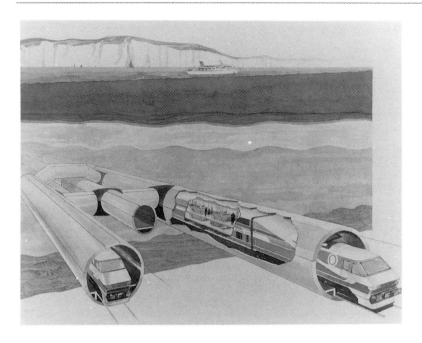

Artist's impression of the 1987 Channel Tunnel project.

owners of the properties chose to leave at any time during that period they would still be offered the market value of the property plus the value of any improvements which had been made in the interim. This was accepted as a very fair arrangement and a number of the villagers sold up and left.

The Tunnel portal was to be at the eastern end of the Folkestone Terminal configuration – at the base of Castle Hill, or Caesar's Camp as it is locally known. The French Tunnel portal was to be at the western end of the Coquelles Terminal at Beuissingue Farm near Sangatte.

THE SAFETY ISSUE

During the passage of the Bill through Parliament, Flexilink changed its tack and mounted an assault on the project's safety. It resorted to a 'shock-horror' video which vividly elaborated what might happen should drivers and passengers remain with their vehicles during the half-hour journey through the Tunnel: stationary cars would burst into flames consuming the luckless occupants and transforming the Tunnel into a blazing inferno. Yet there was no evidence for spontaneous combustion of any road vehicle in the entire history of sea-ferry transport. Nor, for that matter, was there any such evidence relating to the train services which operate through the Swiss Alpine tunnels, where road vehicles have been carried on open flat-wagons since 1924 – a method far less sophisticated than the Eurotunnel system. During the past thirty years such services have been operating through some half a dozen

Alpine tunnels up to 29 kilometres in length, carrying 25 million road vehicles and 75 million passengers, without a single fatality.

If Flexilink could make this issue stick, the throughput of the Tunnel would be drastically reduced. To segregate passengers and drivers from their vehicles would add at least half an hour to the loading and unloading time of each shuttle train. This would be far more serious than a slowing down of the service. The financing of the project was geared to Eurotunnel's traffic and revenue studies and such a downturn in throughput capacity could jeopardise the financial viability of the entire project.

This was the real target of Flexilink's spoiling action, and it has to be said that from 1981 to 1987 the project had been a perennial cock-shy for certain back-bench MPs. Any controversial issue would do, and more often than not it was the Channel Tunnel. The two back-benchers most vociferous in their support of Flexilink were Teddy Taylor (later knighted), Member for Southend East, and, more latterly, Jonathan Aitken. Teddy Taylor, inveterately anti-anything European, could be relied upon to fire a blistering broadside at any issue concerning that 'foreign Continental land mass to the East of Southend', and the Channel Tunnel was an obvious target. Jonathan Aitken's motivation was more obscure, but both men would seize every opportunity to condemn the project. Fortunately they gave voice once too often in support of the wildly exaggerated claims of Flexilink and lost all credibility on the subject of the Tunnel.

However, safety was a real issue. The horrific capsizing of the Townsend Thoresen ferry vessel *Herald of Free Enterprise* on the Dover-to-Zeebrugge route on 6th May, 1987, with 193 lives lost as a result of gross negligence, robbed Flexilink of any voice that might question the Tunnel's safety.

Despite the claims made by Flexilink, there had been no glossing over of the potential hazards of the Tunnel. All safety matters were under the overall supervision of the two Governments; and one of the most important features of the Treaty was the creation of an Intergovernmental Commission (IGC) and a Safety Authority.

The IGC, in brief, would be a joint Government watch-dog committee supervising all matters concerned with the construction and subsequent operation of the Tunnel. The Committee comprised eight members from each country, including at least two members from the Safety Authority. Although a separate committee in its own right, the Safety Authority was to report through the IGC. Its responsibility would cover all aspects of safety, with the power not only to enforce such measures, but to monitor their implementation. Its composition was similar in every respect to the IGC – equal British and French representation and annual rotation of Presidents.

THE FUTURE ASSURED

On 6th April TML moved its offices to Charter House, Ashford, Kent, and opened a training centre for the recruitment of the construction workforce. The Tunnel was basically a low technology project which in no way detracted from its stature as one of the world's greatest engineering projects. Leaving aside 'The Tunnellers', the teams of specialists at the sharp end who would drive the boring machines, the majority of the workforce would span the trades and skills of the entire construction industry. TML regarded the training as an indoctrination to the scale and scope of the work to be undertaken; the surroundings would be totally different from the norm, and great emphasis was placed on safety – of paramount importance in underground workings.

Prime Minister Margaret Thatcher and President Mitterrand ratifying the Franco/British Channel Fixed Link Treaty, Paris 1987.

Meantime, in Paris, the legislative procedure continued to progress. Draft legislation for the Ratification of the Treaty and the Concession Agreement had been passed – after the advice of the *Conseil d'État* on 8th January, 1987 – to the *Conseil des Ministres* (Cabinet), where it was approved on 14th January. The *Assemblée Nationale* unanimously approved the enabling legislation for the project on 23rd April. Following promptly in the wake of this approval, the *Conseil d'État* gave its advice to the Prime Minister who signed the *Décrée de Déclaration d'Utilité Publique* on 6th May; this was duly countersigned by the Minister for Development Administration, Housing, Regional Development and Transport; and published in the *Journal Officiel* on 8th May. Parliamentary procedures were finally completed on 3rd June with the unanimous passing of the Tunnel legislation by *le Senat* and the formal approval of President Mitterrand on 15th June.

TML's first batching plant for the production of concrete tunnel lining segments on the Isle of Grain was completed in June 1987. Over half a million concrete segments, each weighing on average six tons, would be required to line the UK half of the Tunnel. The segmental rings of reinforced concrete would be of the 'expanded type' – a wedge-shaped 'key segment' would 'expand' the lining against the tunnel wall where it was grouted to ensure uniform contact between lining and surrounding chalk. Any ground which was at all suspect would be lined with cast-iron bolted segments for extra strength.

It was also in June that the Robbins 4.5-metre (14.9-foot) boring machine, which had stood like some brooding, mechanical watch-dog on the brink of the abandoned 1975 Tunnel workings at Sangatte, took on a new lease of life. Carefully tended and serviced for over 13 years it was sold to Turkey where it was to be used in the construction of new sewers in Istanbul.

In Britain, however, the political scene was far from encouraging. The General Election was to be held on 11th June, 1987, and Gordian knot of circumstance had begun to tighten. It was an anxious time for Eurotunnel. If the Conservatives were returned for a third term with a workable majority, all would be well, except that Parliament would be in recess for the three weeks preceding the Election date – a loss of time which the Hybrid Bill could ill afford. If a Socialist Government were returned, there would be the likelihood of a Public Enquiry into the project. Such an enquiry would last 30 months at least, and such a delay, whatever the outcome, could spell *finis* to the project for the third time. In the event the Conservatives were returned to Government under the continuing Premiership of Margaret Thatcher with a comfortable majority of 102.

During the month of July the Hybrid Bill moved rapidly towards completion. The Third Reading took place in the House of Lords on 16th July, and on 21st July the Bill was returned to the Commons for its consideration of the Lords' Amendments. It passed into law as The Channel Tunnel Act 1987 on 23rd July. Negotiations between British and French Railways were also concluded after some lengthy and tough bargaining, with Eurotunnel being guaranteed 50% of Tunnel revenues. The Channel Tunnel Usage Agreement was signed on 29th July – a particularly rewarding day as it symbolised the conclusion of Tunnel legislation on both sides of the Channel.

Prime Minister Margaret Thatcher and President Mitterrand met at the Elysée Palace in Paris for the formal exchange of instruments ratifying the Franco/British Channel Fixed Link Treaty. This ceremony was the culmination of a process which had begun in 1984 when the two leaders had first agreed in principle to join the two countries with a fixed link.

In a few words, the Prime Minister, dispelling all doubts that Perfidious Albion would once again withdraw from the brink, succinctly encapsulated 185 years of failed endeavour:

> Too often in the past, pioneering spirits, men of vision and imagination, have been foiled by bureaucracy, narrow minds, or plain fear of the unknown. I hope this time that we can rise above the hesitations of the past; that we can grasp the excitement of this project and the scale of the benefits which it could bring to both our countries and to Europe as a whole . . .

President Mitterrand, ever equal to the occasion, said that he did not think that the feelings of insularity which were so profoundly entrenched in British history and culture would disappear with the realisation of the project. It was more the occasion for a great number of Europeans to get to know and to understand Britain. Nevertheless, Britain would discover France and, through her, Europe:

> . . . and we can safely say that the Continent will cease to be isolated.

With the ratification of the Treaty, the 55-year concession came into force and would remain so until 28th July, 2042.

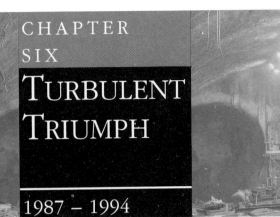

CHAPTER SIX
TURBULENT TRIUMPH
1987 – 1994

NINETEEN EIGHTY-SEVEN

On 25th August, some 50 international banks entered into an agreement with Eurotunnel to underwrite the £5,000 million loan finance, and the process for syndicating the loan to a wider grouping of banks around the world began. Two days later, two groups representing Eurotunnel and the Arranging Banks, supported by teams of technical consultants, set off on a world tour to demonstrate the financial resilience of the project to the representatives of 250 banks who had been invited to participate in a wider syndication[95].

While Eurotunnel was attending to financial matters, TML took over the historic site at Shakespeare Cliff to start preparatory workings. The self-same site where both ill-fated attempts to build the Tunnel had begun would be the setting for all tunnelling operations on the British side. As well as preparing the 1975 access tunnel and a stretch of completed pilot tunnel to receive the first boring machine, work began on access for the main tunnel boring machines. At the base of the cliff, on the Lower Working Site, construction began on a second access tunnel for the right-hand main running tunnel towards France and inland to the Folkestone Terminal. On the Upper Working Site, a vertical shaft was started from the top of the cliff which would provide access for the tunnelling machines for the inland and seaward drives of the left-hand main tunnel. Both the vertical shaft and the second access tunnel gave entry to a large underground working chamber.

---

[95] See Appendix IV.

The French shaft at Sangatte viewed from above (left) and within (below).

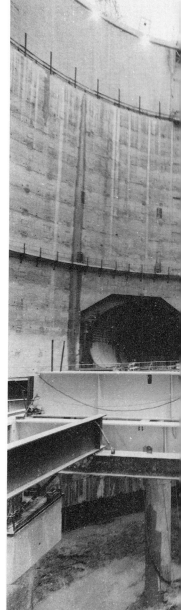

TML was led by John Reeve[96] and François Jolivet[97] as Joint Directors-General; and Philippe Essig[98] and Andrew MacDowall[99] as Chairman and Deputy Chairman respectively. These men, who shared a wealth of experience in engineering, construction and railways, carried the heavy burden of responsibility for the biggest and most important engineering enterprise in the western hemisphere – an enterprise which would tax their combined knowledge and ingenuity to the limit.

---

[96] At the start of his career as a Quantity Surveyor, twenty-year-old John Reeve, when asked what his ambition was, invariably replied: 'To build a tunnel across the Channel to France'. His flippant remark proved prophetic. From the start he had been involved with the construction industry, joining the Costain group in 1967. During the next twenty years, he had developed a multi-disciplined world-wide enterprise in the fields of nuclear and petrochemical engineering. Chairman of three Costain major subsidiaries, he became the driving force in Costain's involvement with the Channel Tunnel, via ECTG and CTG which led to his appointment to TML in 1986.

[97] François Jolivet, born in Lorraine, began his professional career as a design engineer with the Spie Batignolles Group, becoming a Director in 1979. He was responsible, in conjunction with Framatome and Alsthom for the export of French nuclear power plants on a turnkey design construct basis, and was seconded to TML in 1986.

[98] Philippe Essig had a reputation as one of the world's leading specialists on railways and transportation systems, acquired during a career which had involved him with transport methods throughout the world. In 1985 he became Chairman of SNCF, turning a six billion franc deficit in 1984 into a balanced account by 1989.

[99] Andrew MacDowall joined George Wimpey & Co. in 1964, becoming a Director in 1974. He was responsible for major projects in the Middle East and was appointed Group Board Director and Managing Director of Wimpey International & Engineering in 1982. It was during this time that he became involved with the Tunnel project when Tarmac-Wimpey formed CTD 81. He became Deputy Chairman of TML in May 1987.

During the following month, the Calais *Centre d'Information Eurotunnel* was transferred to Sangatte and was sited near the huge shaft – 55 metres (174 feet) in diameter and 60 metres (196.9 feet) deep – which had been sunk to provide access to all tunnel boring operations, both landward- and seaward-drives, on the French side. The sinking of the main shaft would in some measure overcome the influx of water experienced during the 1974/75 project when an access tunnel was constructed to reach the level where the boring machines could start operating. But landward tunnelling from the main shaft, in its ascent to reach the Coquelles Terminal, would still have to pass through the water-bearing, fissured strata of the Middle and Upper Chalk. The angle of incline could be no more than 1 in 90 as electric trains begin to lose traction on steeper gradients. The forward thrust of the French tunnelling machines would be off the Tunnel lining, built in the tailskin of the machine. The spoil would be pumped back in the form of a slurry to a sump at the bottom of the main shaft for onward pumping to Fond Pignon for disposal.

There had been no significant changes from the design and construction of the 1974 project, with the exception that the internal diameter of the pilot tunnel had been increased to 4.8 metres (15.75 feet) and the two main running tunnels to 7.6 metres (25 feet).

The first tunnel boring machine (TBM) for the British seaward-drive of the pilot tunnel towards France was built by James Howden of Glasgow, Scotland, and delivered to TML at Shakespeare Cliff on 1st September, 1987. Dismantled for transportation, it was delivered by road – some 60 lorry-loads – to be reassembled underground at the cutting face. The 5.38-metre (17.7-foot) diameter TBM, costing £3.2 million, was 150 metres (164 yards) long, including the tunnel lining and spoil conveyor sections, with a capability of boring at the rate of eight metres (26.2 feet) per hour. The shield advancement of the British open-face TBM – which would have the facility for closing down the face if the need arose – would be achieved by the use of hydraulic rams and gripper pads acting on the surrounding ground area, or by jacking against the completed lining.

Eleven TBMs were to be used to excavate the 50-kilometre (31-mile) long Tunnel. Six would be required on the British side – three to drive the seaward stretch of the pilot tunnel and two main tunnels, and three to drive the same tunnels inland for a distance of some nine kilometres (5.5 miles) to the terminal at Cheriton. The contract for the two 8.36-metre (27.4-foot) diameter TBMs for the British main tunnel seaward-drive was awarded to Robbins/Markham, with deliveries to the Shakespeare Cliff site scheduled for August and October 1988. James Howden would also build and supply the 8.36-metre-diameter and the second 5.38-metre-diameter TBMs for the British landward-drive.

Five TBMs were required on the French side, as only one 8.36-metre-diameter Mitsubishi TBM would be necessary to drive the two main tunnels the short landward distance of 3.2 kilometres (two miles) to the Coquelles Terminal. Because of the geological fault near the French coastline, all five TBMs were of the closed-face type, designed to work in very wet ground, with the capability to convert to an open-face mode for dry working conditions. The first pilot tunnel boring machine for the seaward-drive would be built by Robbins/Kamatsu for delivery early in 1988. Mitsubishi would build the landward pilot tunnel machine, and Robbins/Kawasaki would build the TBMs for both the main tunnel seaward-drives.

Tunnel workings at Sangatte, 1987.

Discussions and negotiations between Eurotunnel and the European Investment Bank (EIB) had been continuing for several months and were agreeably climaxed with an agreement on 7th September, 1987, whereby the sum of £1 billion would be phased in a series of loans over a six-year period. But a fortnight later a less agreeable situation arose.

Under the headline 'Split Threatens Tunnel Funding', *The Sunday Times* of 20th September reported that it was in possession of a copy of a highly critical letter which had been sent by Eurotunnel to TML. Dated 28th August, the letter, written by Pierre

Durand-Rival, Eurotunnel's Deputy Chief Executive, was a catalogue of criticisms: '. . . slippage in the engineering programme . . . major delays in rolling-stock design . . . total lack of financial information and cost control procedures . . . the present situation is clearly unacceptable and most disturbing . . . '

There was little doubt that the letter had been deliberately 'leaked' to demonstrate to would-be investors in the forthcoming Equity 3 flotation that Eurotunnel was a firm, no-nonsense taskmaster, which would brook no shortfall in the performance of TML's contractual obligations.

Financial observers appeared to be reassured that Eurotunnel was taking its management duties seriously. Only a fortnight later, in an interview with *New Civil Engineer*, Pierre Durand-Rival claimed that he was now satisfied that the detailed points in his letter had been answered: 'I am happy with the control systems which monitor the progress of the project.'

On 1st October, 1987, in preparation for the Equity 3 share issue, which was to be on offer from 16th November to 27th November, 1987, the Eurotunnel Share Information Centre was opened at Winchester House in the City. It was complete with videos and displays, and the star attraction of an 'n'-gauge working model railway layout of the Folkestone Terminal, some 14 metres long and 3.5 metres wide. This was the first time that any share issue had the facility of an Information Centre. On the eve of the campaign launch of Equity 3, Eurotunnel made two new appointments to the Board: Dr. Tony Ridley[100] and Sir Robert Scholey[101].

Then, when all the portents signified that the project was set fair, adversity re-asserted itself in a daunting manner. On 19th October, 1987, Wall Street opened with no signs that it would be anything other than a quiet start to the week's business. It was immediately confronted by an unprecedented deluge of selling that was to rock the very foundations of the international monetary system. As stock markets reeled under the momentum of wave after wave of selling, free market capitalism faced its greatest challenge since the Wall Street crash of '29. But as billions of pounds continued to be wiped off share values around the world, it was recognised that the crisis was far more serious than any historical comparison might indicate.

Throughout its history the Channel Tunnel had weathered many storms and, once again, its future was in grave doubt. For if the Equity 3 issue faltered, or was postponed, the delay could be

---

[100] Then Chairman and Managing Director of London Underground and Chairman of Dockland's Light Railway.
[101] Then Chairman of the British Steel Corporation and President of the Association of European Steel Manufacturers.

disastrous. The shock-waves of Black Monday were turbulent enough to threaten Eurotunnel's relationship with the international banking consortium, resulting in a loss of forbearance and withdrawal of support, there being no lack of less uncertain opportunities in which to invest new money.

On 4th November, however, the Credit Agreement for £5 billion Project Finance Credit Facilities was signed simultaneously in London and Paris by Eurotunnel Finance Limited, Eurotunnel SA and the original 50 international banks, whose commitment in principle had played such a decisive role in the awarding of the Mandate. It was also signed by a further 150 international banks who had subsequently participated in the wider syndication of the loan arrangements[102]. Unfortunately, Eurotunnel still had to raise the final balance – £770 million – of its £1 billion of public equity, this being the risk capital which it would have to spend before it could draw upon the loan capital.

Gradually the money markets began a shaky recovery and it was into this post-crash aftermath, still twitchy and unsettled, that Eurotunnel was to launch its all important final tranche of equity. Needless to say, the crash had shaken the investors' confidence, particularly those who had taken their first faltering steps, encouraged by the British Government's popular capitalist programme, into the realm of stocks and shares.

André Bénard (above) and Alastair Morton (below), Eurotunnel's Co-Chairmen.

Black Monday had also been a traumatic experience for the traditional financial institutions, and more importantly the underwriters, whose support against the backdrop of a nervous market would be vital to the success of the issue. A less opportune moment to venture into the market place could scarcely be imagined.

However, the underwriting syndicates had recovered sufficiently to be in business as usual on 16th November when the Equity Underwriting Agreement was signed between Eurotunnel PLC, Eurotunnel SA, and the Underwriting Syndicate. The outcome was no longer in doubt. The vital £770 million was assured. Eurotunnel had cleared the final hurdle – not handsomely, but creditably, given prevailing stock market conditions.

The main prospectus was issued on the same day.

---

[102] See Appendix IV.

Only £353 million of the total would be placed through the London Stock Exchange. A similar sum would be placed through the Paris Bourse, and the balance of £64 million in Belgium, Japan, Saudi Arabia, and the United States.

The shares carried certain investor inducements. There were warrants – one for each ten shares – which could be used to purchase future shares at a preferential price and in addition there were short-, medium-, and long-term travel privileges. The purchase of 100 shares entitled the owner to one free return trip through the Tunnel during the first year of operation; 500 shares carried one free trip per year for the first ten years of operation; 1,000, two free trips a year for the period of the Concession; and 1,500 entitled the purchaser the right to unlimited free travel through the Tunnel for the entire Concession. These travel privileges, however, were vested only in the original purchaser; they were not transferable, and would expire on the transfer or re-sale of the shareholding. Shortly after the closing date of the issue, it was announced that there had been a shortfall of 20% in the UK issue and some 15% in France. This shortfall would be reflected in the value of the shares when trading began.

At a cost of £68 million, Equity 3 was arguably the most expensive stock market flotation ever staged. There was a duplication of costs because the issue was offered internationally and floated simultaneously on both sides of the Channel. Fees and commissions to financial advisers, merchant banks, underwriters, and receiving banks accounted for £40 million; advertising some £7 million; with legal fees making up the balance. There was too much at stake, however, to risk the issue going by default – it was no time to be counting the cost.

One factor which may have affected the shortfall in the take-up of the shares was that the market research undertaken prior to the share issue was misleading. It indicated that 574,000 UK investors were 'certain' to take part in the share offer – whereas in the end only 112,000 applied. Although this number was quite respectable, the relationship between merchant bankers S. G. Warburg, one of Eurotunnel's leading advisers, and the marketing consultants became less than cordial. The discrepancy between their 'guesswork' and the end result could have led to an easing of the marketing effort before the completion of the flotation, as a vast share register at the level indicated by the market research would have been costly to maintain[103].

---

[103] Even so, Warburg Securities' case for the investment merits of the British issue was persuasive. This predicted that, although the shares would not be paying a dividend before 1994, their value would increase sevenfold by the time the Tunnel opened in 1993; this was because the project had little by way of assets at the time of issue, but would have an asset value of some £6 billion on completion.

Compared to earlier privatisation issues, Eurotunnel's publicity campaign during the weeks preceding the Equity 3 flotation was a somewhat pedestrian affair. There were possibly extenuating circumstances as to why a campaign more worthy of the 'greatest engineering project of the century' did not emerge. Not least being that before the creative juices could flow, any newcomer to the project was faced with a very steep learning curve in order to reach a full understanding and appreciation of its many facets and implications. There was never enough time for this, or for consideration and selectivity – ideas were steam-rollered through for want of something better in order to meet deadlines, with decisions at times being made by those lacking a real perception of what was required. This produced one of the worst and one of the best examples of the creative art, with the run-of-the-mill sandwiched somewhere in between.

The worst advertisement was a full page of heavy type divided down the centre by a zigzag of white space: *THE CHANNEL TUNNEL IS THE GREATEST THING SINCE MOSES PARTED THE WATERS OF THE RED SEA.* This piece of trivia did nothing for the project except to suggest that the only way there would ever be a Channel Tunnel was by divine intervention.

At the other end of the scale was an exceptional example of the creative art. Against a panoramic seascape of the Channel, the White Cliffs of Dover to the left and the French coastline to the right, stood the words: *FROM A TO B AVOIDING SEA.* Underneath, in smaller type, were the words: *A breakthrough for Britain.* Succinct, memorable, and well deserving of its award-winning recognition.

The French perception, by contrast, was just right. Their publicity campaign, conducted by the Alice Agency, unerringly reflected the public's attitude. True, they had more latitude as there was little or no militant opposition to the project and possibly some of their publicity would not have been condoned under the rules governing share issues in Britain. Nonetheless, their style was prestigious and grandiose – it was 'the greatest project of the century' and 'shareholders would be making a rendezvous with history'. The occasional advertisement to the effect that 'The Channel Tunnel was an investment in the future prosperity of the Nation' would not have come amiss in Britain.

The pilot tunnel TBM in position at the Shakespeare Cliff seaward heading was now ready to start its commissioning trials. On 1st December, 1987, witnessed by men facing the ultimate challenge of their careers, the huge cutting head made its first revolution. This time it was the beginning of the end. Twice before a similar scene had been enacted, only to leave those taking part to savour the bitter bile of defeat and disillusionment. Now, political relations between Britain and France had never been more cordial; the legislative instruments of Parliaments and the Treaty were in place; and the private finance had been safely gathered in, sufficient to carry the project through to completion. In fact, TML was so confident of its ability to complete the project on time that it had named a date for the Tunnel's opening – 15th May, 1993.

It had been a year of remarkable accomplishment, concluding with the attaining of the ultimate goal and, in so doing, rounding off a chapter of 186 years of history. The year also marked the passing from the scene of Commander Christopher Powell[104], and Colin Stannard[105], both of whom contributed much to the project's successful conclusion.

NINETEEN EIGHTY-EIGHT

On 29th January, President Mitterrand, escorted by André Bénard and Alastair Morton, visited TML's access shaft at Sangatte. The President witnessed the assembly and installation of the Robbins/Komatsu TBM for the pilot tunnel seaward-drive. He was generous in his praise of the achievements thus far, and emphasised the project's importance to the future of Europe and France.

The second access tunnel at Shakespeare Cliff under construction, 1987.

[104] Commander Christopher Powell, RN, Secretary of the Parliamentary (All-Party) Channel Tunnel Group since 1929, retired on 20th July, 1987. He was the well-regarded and much-respected doyen of Parliamentary Consultants who had diligently maintained the political profile of the project for over half a century.

[105] Colin Stannard had initially been seconded from the National Westminster Bank, where he was one of the original authors of the 'Fixed Channel Link Report' of May 1984. He had also been, together with Jean Renault of Spie Batignolles, Joint Managing Director of Eurotunnel. Subsequently, as Managing Director (Commercial), he was one of the leaders of the British team whose successful submission resulted in the award of the Mandate, and he played a major role in securing the commitment whereby the international consortium of banks were to pledge the required loan capital. Major contributions were also made by the groups of officials in the British and French Ministries of Transport who steered the complex legal and intergovernmental procedures through. Two names deserve special mention on the British side, having been associated with the project since the 1970s: John Noulton, now Under-Secretary of the Transport Directorate and one of the alternate Presidents of the project's Intergovernmental Commission; and Ted Glover, a civil engineer with the Department of Transport, whose technical knowledge of the project was second to none. Both men were steeped in fixed-link lore and their long-term memory was invaluable.

A week later, on 5th February, Prime Minister Margaret Thatcher, escorted by Eurotunnel's Joint Chairmen, visited TML's construction site at Shakespeare Cliff where they were conducted on a tour of the workings by John Reeve. The Prime Minister sat at the controls of the Howden boring machine – which was still undergoing commissioning trials – and expressed her delight with the progress that Eurotunnel, and its contractors, had made:

> This is an exciting project which is bringing great opportunities
> to the people of Kent and to the whole of the United Kingdom.

In the latter part of 1987, David Mitchell had been given a decidedly rough ride at a meeting with the French Ministry of Transport. On his return he immediately directed BR to implement a study for a new Tunnel-dedicated, high-speed track from London to the Folkestone Terminal.

The Kawasaki/Robbins TBM, 1988.

Beaumont's 1882 tunnel further uncovered as the main boring machine bisected it.

Throughout the early 1980s BR had opted for a 'no new line' strategy, no doubt bearing in mind the public outcry which greeted its 1974 high-speed track proposals. To use the West London line via Olympia and West Brompton would connect main lines from the North to the South-East network, enabling BR to operate a service from, say, Glasgow to Milan virtually non-stop. BR was confident that it could serve the country's Continental freight and passenger requirements into the early part of the next century. In the first year of Tunnel operation BR expected to triple its current Continental-bound freight of 2.9 million tonnes. This alone would achieve a major environmental plus, being the equivalent of removing 1,500 38-tonne juggernauts from Britain's roads every day.

David Mitchell's directive was highlighted during Eurotunnel's AGM, held on 18th April, 1988, simultaneously in London and Paris. Both André Bénard and Alastair Morton made strong pleas that a British integrated transport plan was essential if the benefits of the Tunnel were to be fully realised. BR's high-speed track was back on the agenda[106].

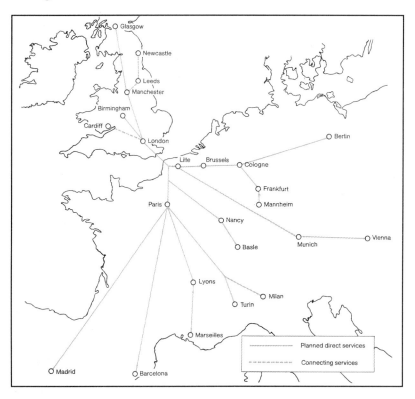

The 1974 high-speed track proposals.

On 12th May the second boring machine was delivered to Sangatte. The 300-tonne TBM, built by Mitsubishi, would be boring the landward pilot tunnel towards the Coquelles Terminal, a distance of 3.2 kilometres (two miles). It was scheduled to finish its commissioning trials by July and to complete this section of the pilot tunnel by the Spring of 1989. Boring operations in general, however, were not proceeding as planned. The French pilot tunnel

---

[106] Sadly, it was to remain 'on the agenda' until well after the Tunnel's official opening.

seaward-drive TBM was still being commissioned, and the British pilot tunnel seaward-drive was only 1.2 kilometres – less than a mile – out from Shakespeare Cliff. A combination of bedding-down problems and difficult ground conditions had hampered progress.

On 4th July Eurotunnel relocated to the newly-completed Plaza Complex at Victoria Station. Here, Alastair Morton continued his campaign to press home the need for investment in a dedicated rail connection through, or around, London. A few days later, on 14th July, BR revealed its answer to his plea with the long-awaited proposals for a high-speed route from London to the terminal at Cheriton. The study proposed three possible routes through Kent, with the options of Kings Cross, Stratford East and White City for the second London terminal, Waterloo being the first. The cost of the proposed new routes ranged from £725 million to £1,200 million. The preferred King's Cross route, subject to commercial viability and environmental considerations, would take eight to ten years to complete and would not be in service until 1998 at the very earliest – some five years after the Tunnel was in operation. The new high-speed link would reduce the journey times from London to Paris to 2 hours 30 minutes, and from London to Brussels to 2 hours 20 minutes.

It was also during July that an incident took place which highlighted the fact that anti-Tunnel factions were still at work. In the early hours of 20th July, one of the small-gauge locomotives used for hauling the spoil from the working face at Shakespeare Cliff developed a faulty electrical connection which sparked a small fire. The driver had put out the fire within a matter of matters using a hand extinguisher, but as smoke and acrid fumes had been given off, the workforce was evacuated from the immediate area. In accordance with the strict safety procedures, the local fire brigade was notified and five fire engines, two tenders and 50 firemen arrived on the Shakespeare Cliff site ten minutes after the fire had been extinguished. After checking the area, the Kent Fire Brigade Chief agreed that it was a minor incident, that there was no flaw in TML's safety procedures and he approved the return to work at 7.45 a.m. It could have been assumed that this was the end of a minor episode. But not so!

Later in the day, despite the fact that Eurotunnel had issued a statement giving the true facts, several provincial evening papers carried the story of how '. . . eleven fire brigades fought for two hours to bring the blaze under control.' The following morning at least one local radio station reported that 'millions of pounds worth of damage had been caused by the blaze.'

The Transport & General Workers' Union and, as would be

expected, the marine officers' union NUMAST, had been strongly opposed to a Channel Tunnel whilst the railway unions NUR and ASLEF, and the construction workers UCATT were in favour. Latterly, however, the Fire Brigades Union FBU had been particularly militant. As recently as June, Eurotunnel had cause publicly to rebuke Ken Cameron, General Secretary of the FBU, for unsubstantiated criticism of the Channel Tunnel during the public enquiry into the King's Cross London Underground Station fire disaster. This was by no means a reflection on the amicable relations and co-operation which existed between the Kent Fire Brigade and TML; but such a distorted version of a minor incident could only be the result of deliberate mischief-making. It had obviously been syndicated via a news agency and was to prove typical of what was to become long-term 'knocking copy'.

Marine works at Shakespeare Cliff – constucting new sea wall for disposal of spoil from excavations.

Towards the end of July serious problems were developing on the tunnelling front. The construction schedule was divided into key 'milestones' and, having met the deadline for Milestone 1, TML had failed to meet deadlines for both Milestones 2 and 3. The next target was Milestone 4, which called for the completion of five kilometres of the British pilot tunnel and two kilometres of the French pilot tunnel by 1st November, 1988. TML was only 1.6 kilometres out from the British coastline and somewhat less than 200 metres out from the French side. It now seemed likely that TML would fail to achieve Milestone 4 within the allotted time and would be liable to incur severe financial penalties.

In a revealing interview with John Allen, Editor-in-Chief of *Construction News*, John Reeve, Joint Director-General of TML, gave an insight into the contributing factors which had directly affected the construction schedule. At the time of the launch of Equity 3, in the opinion of the *Maître d'Oeuvre*[107], the project was already running some three months late. The delays with Equity 2 and 3 had imposed severe cash restraints on both TML and Eurotunnel, with the result that TML was having to work on a month-to-month basis as to commitment and expenditure. These cash constraints led to delays in commitment to design studies in other areas, as any available funds were concentrated on the all-important details of Tunnel design. Also, the passing of the Channel Tunnel Bill and Royal Assent had come two months later than had been expected. This in turn was critical, as TML could not claim right of access to the working site at Shakespeare Cliff until these procedures had been completed. 'We have to recover that time', said John Reeve, who went on to say that once the tunnelling teams had settled in, the job should run fairly smoothly and, having re-checked the geology, the contractors were entirely satisfied that the conditions would hold few, if any, surprises.

Following the publication of BR's studies on 14th July, Alastair Morton was extremely unhappy with the prospect of the new line not being ready before 1998. On 12th August he stated his views in a letter to Sir Robert Reid, Chairman of British Rail. He stressed that the line would be 'very necessary' by 1995, two years after the Tunnel was open – a conclusion which was based, he said, on early indications of Eurotunnel's 1988 up-date of traffic forecasts. Morton further confirmed his confidence that private sector risk capital could be harnessed for substantial elements of the new high-speed

107 A French term for the traditional independent Consulting Engineer. The role of the *Maître d'Oeuvre* was undertaken by a group of independent British and French consulting engineers, comprising W. S. Atkins & Partners and SETEC of France. The *Maître d'Oeuvre* was retained by Eurotunnel, but also monitored progress for the banks and the Intergovernmental Commission.

line – a process which Eurotunnel would support as catalyst or promoter. The text of the letter to Sir Robert was released to the press on 14th August – a tactic on Eurotunnel's part which was becoming all too familiar.

At the end of August Morton turned his attention to another matter which was causing him concern. In an astonishing outburst, he vented his frustrations on the ten parent companies of the British and French consortium which made up the TML construction company. He roundly accused all those at board room level of lack of interest and support for the project, complaining of their failure to devote sufficient management expertise to the Tunnel. He said:

> There's a tendency among them to say: 'We won the contract and supported Eurotunnel's finances, so now we can go off and do other projects.'

On the face of it, the facts seemed to support Morton. During the previous month, tunnelling on the British side had only advanced 200 metres, and on the French side only a few metres. Eurotunnel therefore served formal notice on TML under Clause 46 of the Construction Contract, whereby Eurotunnel, in the event that it considered the contractor's rate of progress to be too slow in any stated activity, could require TML to take steps – with Eurotunnel's approval – to expedite progress on that activity.

Apart from the problems inevitably associated with the start of a project of this scale, there was a specific reason why tunnelling was behind schedule. The French TBM was delivered several months late, and the bad ground conditions caused its performance to fall far short of its design specification. As one French worker on the site at Sangatte said: 'The English may be in trouble – we are in worse trouble here.'

Similarly, at Shakespeare Cliff, the British had struck an unanticipated damp patch. Although the project was now three months behind schedule, this problem was not yet serious; but prudent and efficient management would be essential if the project were to be completed on time.

No doubt aggrieved at being bound to silence and having, by association, learned some bad habits, TML now leaked Eurotunnel's warning letter to the news agencies. The letter contained the following passage:

> The issue of the formal warning cannot remain a secret in the current climate of concern among analysts and journalists. There is too much half-complete awareness of tunnelling details in the media already, and we are under pressure from the City ... We propose to add the minimum by way of comment, and I wish to confirm to you that we oppose any comment from

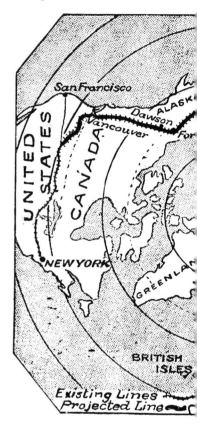

The 1890s vision of England as the centre of a European rail network.

TML beyond something along the lines in quotes:

'TML has received Eurotunnel's notice and is giving it every attention.'

Unless, of course, you wish to add that you are confident of completion and opening in May 1993. We would be happy to hear that said.

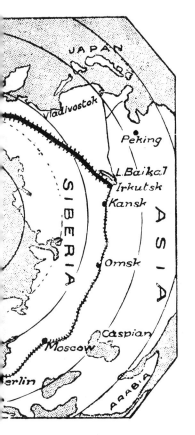

This fostered speculation as to motives, and conjecture as to the effect that the less-than-satisfactory progress of construction would have on the release of funds. The first call on the lending banks was due to be made by Eurotunnel at the beginning of November. However, provided that the balance between costs and forecast revenues did not disturb the financial ratios employed to evaluate the Tunnel's progress, release of funds should be prompt. A total of some 14 months delay would have to elapse before the international consortium of banks would contemplate withholding payments.

While a no doubt acrimonious dialogue was being held behind the scenes, outwardly there was a temporary lull in the dispute, during which time Eurotunnel's £2 million purpose-built Exhibition Centre on St. Martin's Plain, Cheriton, was officially opened on 19th September by Lord Young, Secretary of State for Trade and Industry. The Exhibition was designed to entertain and inform all ages via displays and working and static models, including a walk-in reconstruction of 1880 tunnelling conditions, a full-scale reproduction of a double-decked, car-carrying shuttle, and probably the world's largest 'n'-gauge working railway models of the Folkestone and Coquelles Terminals. In addition to the exhibition, there was a French-style café, and outside was a 70-foot-high observation tower overlooking the Folkestone Terminal site. It was hoped that the Exhibition would attract some 300,000 visitors a year.

By the beginning of October the interim financial statement published by Eurotunnel did much to ease relations with both TML and the banking consortium. It was a mixture of good and bad news, with the good prevailing. There had been a major overspend of £353 million, but the 7% increase in costs was offset by new traffic and revenue projections, which forecasted a 6% increase in revenues by 1993-94; a 10% increase in the year 2003; and 16% in 2013, a result which would enable Eurotunnel to meet the lending banks' criteria of viability. The indications were that TML was heading for a £110 million deficit in the total target cost – any such loss to be split on a ratio of 70:30 between Eurotunnel and TML.

Eurotunnel now claimed to have identified the cause of the current difficulties, but it was too early to see whether TML would be able to bring the project back on schedule. The TBMs on the

British and French pilot tunnel seaward-drives had both been plagued by breakdowns and had to be re-engineered underground. In his presentation of the financial report, Alastair Morton was both encouraging and bluntly matter-of-fact:

> We don't have a tunnelling problem. We have an equipment and management problem. Bad ground is not to blame for the delays.

He was referring to the success of the second French TBM on the landward-drive of the pilot tunnel towards the Coquelles Terminal. This Mitsubishi machine was already 130 metres ahead of schedule, having completed 580 metres since it started in July. Only a few metres away, tunnelling through similar ground, the Robbins/Komatsu TBM, on the seaward pilot tunnel, had advanced just 200 metres since March, having been at a standstill for most of the Summer. Morton, however, appeared to go out of his way to be less aggressive. In fact, he praised TML's efforts:

> Improvements are already visible. The UK service tunnel advance of 640 metres in the six weeks to 2 October compares with 1000 metres in the three months before that.

Running tunnel south seaward drive TBM under construction at Markhams in Chesterfield.

Referring to his swingeing attack on TML in August, Morton added:

> The most effective backside kicking has been within TML,
> which has put pressure on the manufacturers of the boring
> machines to make them work properly.

However, he quickly lapsed into his more usual pragmatic tone on summing up with a terse reminder:

> If TML misses milestone 4 by three months, it will have suffered
> by more than £10 million in withheld payments.

Meanwhile, political pressure for private capital to play a substantial role in BR's proposed high-speed track was growing. BR recognised that direct fast access for its passenger and freight services throughout the Continental rail network was a golden opportunity to revitalise its commercial potential well into the twenty-first century. It might not be averse to an alliance of public and private capital, provided that BR was the dominant partner. However, Paul Channon – now Secretary of State for Transport – was a hard-liner on the Government's privatisation programme. Were political opinion to go firm on the solution of a wholly or partially privately-financed new line in operation by 1996, BR would face a further problem. Its £500 million investment commitment for the completion of the 'no high-speed line' alternative in time for the 1993 opening of the Tunnel would prove to be money wasted, as it would not be possible to achieve the rate of return on investment demanded by the Treasury if the new line became operational before 1998.

The groundswell of opposition to the proposed rail link through Kent also began to make its impact. BR's lack of expertise in the art of public consultation left it wide open to scorn on the occasion when scores of householders, who had been assured that the railway would by-pass their homes, woke up one morning to find that they were in the direct line of the route and that their houses would be demolished. It turned out that a member of BR's project management team had taken work home and used his wife's greaseproof paper to trace the route on the kitchen table. The paper had slipped and sent the proposed route on a demolition course slap through the houses. A prompt admission of this human error by BR would probably have been acceptable, but it was left to the miscreant to own up to his mistake publicly, and to describe in detail the manner of its making!

BR's 1988 proposals had the rare effect of mobilising opposition across the entire political spectrum. Protesters in South London were particularly opposed to BR's preferred option of King's Cross, where the Tunnel passenger terminal was to be built below surface level. Logically, this choice had much to commend it.

It offered excellent connections to the national rail network and to transport interchange facilities. International trains could connect with a wide range of direct services: Gatwick and Luton airports, Hertfordshire, Cambridgeshire, Bedfordshire, East Midlands and Humberside. It also offered easy interchange between international, domestic and metropolitan underground services, being served by five underground routes, with a Piccadilly Line route direct to London's Heathrow Airport, and South-East's Thameslink trains. There were direct rail links to the East Coast main line, with through services to the North-East and Scotland; and it would not be difficult to provide a direct link with the West Coast main line serving the West Midlands and the North-West.

In the same year militant opposition had emerged in France for the first time over the routing of the new *TGV-Nord* line, which would connect, via Lille, with the Coquelles Terminal. The French protest, however, was of a different order – they were angry not because the TGV would run through a particular area, but because it would bypass it! Amiens – capital of the Northern French province of Picardy, immortalised in song, and the notorious 'killing ground' of World War I and countless other battles since the 12th century – became another kind of battle-field.

The *Amiennois*[108] argued that to run the line through Amiens would not only save £300 million, but would cut the journey time from Paris to the Coquelles Terminal by 22 minutes. Having failed to make any progress by democratic ways and means, they now opted for mean ways. With a Gallic shrug of the shoulders they bought up prime Picardy farmland along the route chosen by SNCF, and re-sold it to the general public in one-metre-square plots for FF10 (about £1.00 at that time) each. Their strategy was that when French Railways came to purchase the land for the TGV line it would be faced with the task of tracking down thousands of smallholders and, having done so, with a mountain of paperwork to process. This administrative nightmare would take at least six months to unravel – thus postponing the completion of the new line beyond the opening of the Tunnel in 1993.

NINETEEN EIGHTY-NINE

On 20th January, 1989, the Tunnel claimed its first casualty. Andrew McKenna, a nineteen-year-old engineer's assistant, was walking beside the rail tracks half a mile down the pilot tunnel, when he was crushed to death beneath the wheels of

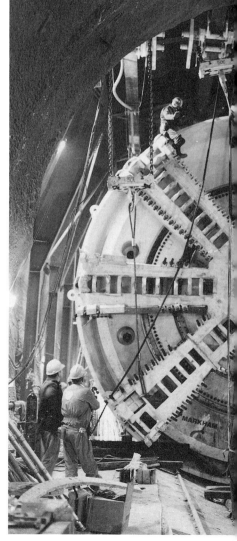

---

[108] Amiens Association.

one of the mini-locos hauling the spoil from the Tunnel. By an extraordinary quirk of circumstance, it was the self-same type of accident which befell Henry Beckingham, the first victim of the original attempt to build the Tunnel in the 1880s. Beckingham was the engineer in charge of boring operations in the pilot tunnel, and was crushed by a wagon as it was carrying spoil from the tunnel face. The construction industry is rated as being the most dangerous of occupations – never more so than when underground. Regrettably, before the project was completed, Andrew McKenna would not be the only one to pay the ultimate price.

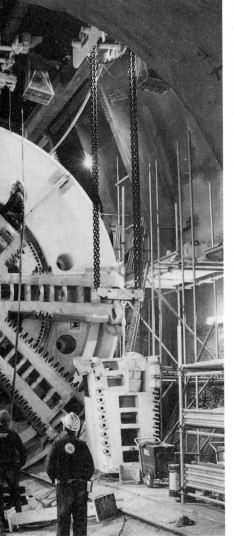

Maintenance being carried out on the cutting head of a TBM.

Towards the end of January, the opposition to BR's proposed routes for the high-speed line assumed dramatic proportions. Around 10,000 people staged a mass protest march through the streets of Maidstone, ending at the steps of Kent County Hall, where a petition of some 14,000 signatures was presented to Tony Hart, leader of the Conservative-controlled Council. The protest march, representing 76 action groups along the proposed routes, was led by five Kent Tory MPs. With the EC 'single market' coming into effect in 1992, the Council, although recognising the need for a new track, rejected BR's three proposals and called for a comprehensive re-think. Andrew Rowe, Conservative Member for Mid-Kent and President of the North Downs Rail Concern Group, backed the KCC's proposal that BR should go back and do its homework properly. He commented:

> This time I hope the plans will be produced from a design table and not a kitchen table.

While all these issues affecting the operational aspects of the Tunnel were being studied, evaluated and debated, morale was running at an all-time high, with record progress being achieved at both Sangatte and Shakespeare Cliff. The French pilot tunnel seaward-drive had completed almost a kilometre in the badly-fissured Middle Chalk area and had broken through into the Lower Chalk which would continue for the rest of the way. The landward pilot tunnel towards the Coquelles Terminal area was still ahead of schedule, having achieved 1,300 metres. On the British side, Tunnel boring speeds were well up to target rates; 230 metres had been attained during the best week so far; and in the last three weeks of January, 600 metres had been attained, with a record 37 metres in one day. The seaward pilot tunnel-drive had passed the five-kilometre mark and the inland drive had already reached three-quarters of a kilometre along its route towards the Folkestone Terminal area.

At these speeds, the prospects of TML eliminating lost time and breaking through as planned were encouraging. However, the blunt exchanges of October 1988 between Eurotunnel and TML still rankled. *Construction News*, reporting the improvement in progress, commented:

> John Winton, TML's construction director, is unhappy at suggestions that improved performance was in any way due to recent harsh words from Eurotunnel chairman, Alastair Morton. It is unbelievable that Alastair Morton considers that the increased production in the last six weeks has been due to his comments.

However, John Reeve, TML's joint Director-General, provided an ebullient insight into working relations. Asked how he got on with his French colleagues, his initial response, was 'About as easily as they get on with me', and he went on to say:

> Of course there are difficulties, but nothing insurmountable. There is of course the language difference, but they are very talented in this respect, whereas we're just abysmal. There is also the way they think – I can't explain what it is, but our minds seem to work differently. It must be a national characteristic – one thing for sure, it's not bloody-mindedness! We can sit around a table with our opposite numbers and within minutes reach absolute agreement on any objective, and in as many minutes both sides will arrive at precisely opposite means of achieving it. But we work it out – there is no room for compromise on this job: it has to be right.

By this time, the full impact of BR's proposals had given rise to a highly-charged political situation – exacerbated by three public enquiries into the causes of railway accidents[109] within the space of three months. Nevertheless, on 8th March BR published a report which attempted to take into account the opinions and criticisms which its original proposals had provoked. The preferred route was:

> From King's Cross in tunnel, ascending to a sub-surface junction at Warwick Gardens near Peckham Rye and descending into twin bore tunnels all the way to Swanley. From there, a new track would be laid alongside existing tracks to South Darenth through a new North Downs tunnel, before crossing the Medway at Halling on a new viaduct to Detling. New tracks alongside the M20 and its Maidstone/Ashford extension would enter a new Ashford tunnel and emerge to run through Ashford's new international station and alongside existing tracks to the Folkestone Tunnel Terminal. Total distance 109 kilometres (68 miles); two-thirds of the route in tunnel or below

---

[109] Clapham 12th December, 1988; Purley 4th March, 1989; and Glasgow 6th March, 1989. There were a number of fatalities and many injured.

surface level. Only 24 kilometres (15 miles) would be a new surface corridor.

This route was close to the original Route 2 published in July 1988, but the maximum speed had been reduced and would range between 160 and 225 km.p.h. (100-140 m.p.h.) over different sections of the route. The planned journey time of 40 minutes between King's Cross and the terminal would, however, be maintained, as the speed of 100 m.p.h. through the London tunnel would be much higher than could be achieved on the surface through London's suburbs.

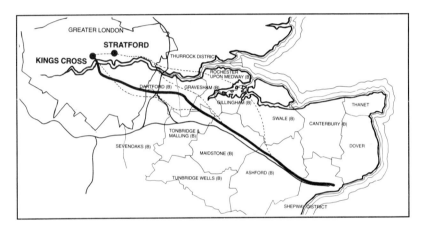

The proposed Channel Tunnel Rail Link, 1989.

The preferred route would save some 5,000 homes. Only 100 houses would have to be demolished and parts of Kent's most attractive villages would remain intact. While the news would be good for some, it would come as a shock to others, who would have learned for the first time that they were affected. Even so, compensation would be at the market value of the properties.

The route would bring an additional benefit to the sorely-tried Kent commuter. Contrary to some beliefs, it would be possible to mix 125-m.p.h.-commuter trains with 140-m.p.h.-international passenger traffic, and there would be connections to the high-speed line to the west of Ashford for fast commuter trains. This would effectively reduce the journey time from Ashford to London from 65 minutes to 35 minutes. Richard Hope, the Editor of the *Railway Gazette International,* who since the early 1960s had so ably advocated the construction of a Tunnel – whilst commiserating with home owners whose property lay in the path of the new line – said that the boost to commuting services in Kent would inevitably send property values soaring.

All very attractive, but at what price to the commuter? The total cost of BR's new environmentally-friendly package had increased the original estimate of £1.2 billion to £1.7 billion. The extra £500 million represented the cost of extensive tunnelling and cuttings, which – if private investment were sought to finance, construct and own the new line – could, with interest and inflation, result in an out-turn cost of over £3 billion. At this level of investment, any hopes held by BR and SNCF of attracting passengers away from the short-haul airline triangle London/Paris/Brussels with cheaper and quicker city centre to city centre journey times, would begin to fade. Quicker maybe – but cheaper?

Reactions to BR's new proposals were mixed. Kent County Council gave it a guarded welcome. Peter Snape, the Labour Party Transport spokesman, who had campaigned so effectively for the Tunnel from 1985 onwards, said in an interview with *The Daily Telegraph* of 8th March:

> The first question is, who will pick up the extra bill for the environmental safeguards? It is the Government's responsibility, not BR's, and there is a great worry it will come out of BR's overstretched finances.

If the new high-speed line was to be commercially viable – whether built, owned and operated by BR, or by a private investment consortium – there was clearly a case for funds being granted for environmental measures. Whether this would come from the British Government or from an EC Regional Grant was problematical. There is a clear-cut binding clause in the Channel Tunnel Act which states specifically that:

> . . . no grants should be made . . . for the purpose of the provision, improvement or development of international railway services.

Any funding from the EC would in fact be British taxpayers' money, already contributed to the EC, being re-cycled back to source. However, Kent's Tory MPs did not see it that way. As Roger Gale, MP for Thanet North, expressed it:

> There is a fundamental difference between construction money and money for environmental protection. No one is seeking money for the Tunnel or the rail link. But there are many of us looking for strong environmental measures.

There was a grain of comfort for BR. Despite an earlier intervention of the Prime Minister[110], BR had secured an agreement with

---

[110] Prime Minister Margaret Thatcher had ruled out on-board train checks. *The Sunday Times* of 9th October, 1988, reported that this decision infuriated French rail officials; they complained that this would add up to half an hour to journey times and make European expresses uncompetitive with the airlines. 'Doesn't she realise that "V" in "TGV" means speed?', said one official.

Customs and Excise in respect of on-board passport and customs checks, which would enable BR to extend fast, international, through-rail services to principal towns throughout the country. Negotiations were continuing on the accommodation for check, search and detention facilities, which had been reduced from the original requirement by 75%.

On 3rd April Eurotunnel issued a further financial and progress bulletin which marginally compensated for the fact that the opening date of the Tunnel had been put back a month to 15th June, 1993. The original cost estimate for the project had risen from £4.88 billion to £5.45 billion. The good news was that TML had made such progress as to be almost back on target. It had forged ahead and in the first three months of 1989 had completed eight kilometres (five miles) – a tunnelling rate three times faster than the previous year. Both terminal areas had been cleared and partly consolidated, and if TML completed and handed over the project by the new completion date, it stood to earn a bonus of £100 million. As Morton put it:

> The next few months will be crucial for tunnelling rates. But we are now pulling together with the contractors, and I am confident that we have sorted out the teething problems.

'We are now pulling together with the contractors . . .' This was perhaps the best news of all. Nothing is more destructive than a divisive relationship between client and contractor. Not only does it mar the public profile of a privately-funded venture, but it undermines the morale of the workforce. In times of crisis, the ability to 'bite on the bullet' is a far more effective prophylactic than the one-sided venting of recriminations in public. It was hoped that the unfortunate breakdown in relations between Eurotunnel and TML during the latter part of 1988 was to be relegated to the past.

Now it was revealed that early in 1989 a Joint Accord had been signed between Eurotunnel and TML which had come into effect in March. It appreciated each party's difficulties, reconciled past squabbles and recognised that joint co-operation between the two companies was essential if the project was to be completed on time.

And if endorsement was needed to cement the new relationship, on 27th April, 1989, the Mitsubishi TBM, driving the 3.2-kilometre (two-mile) inland stretch of pilot tunnel from Sangatte, broke through on the terminal site at Coquelles. Weeks ahead of schedule, it also established a new world record in its category of 886 metres in one month.

The Joint Accord provided one month's extension of the

completion date for TML; increased the overall cost of the project by some £200 million; and settled all outstanding previously disputed payments. A condition of the Accord was that TML improve the quality of its management. TML acceded to this demand, and slotted a new management tier into its decision-making process, but the price was a heavy one. Both John Reeve and his French counterpart François Jolivet, Co-Directors-General, declared themselves 'profoundly disillusioned' with the new management structure, and resigned in June. They believed TML's existing management to be both effective and efficient. It had overcome the difficult problems of the previous year and had successfully demonstrated its calibre both of the tunnelling workforce and of the management by a commendable rate of progress – 25 kilometres (12.75) miles of tunnelling.

In the re-shuffle Philippe Essig became Chairman of the new TML board, composed of ten senior directors drawn from TML's parent companies. Andrew MacDowall became Deputy Chairman[111]. Klaus Van Der Lee[112], who had been responsible for tunnelling operations during the 1974 attempt, became Technical Director. John Noulton, who had been associated with Tunnel matters since the 1970s with the Department of Transport and latterly as alternating President of the Intergovernmental Commission, resigned from the Civil Service to become Director of Administration. Jacques Thibonnier, with extensive experience in major civil engineering, mechanical and electrical works, was appointed Director of Engineering and Transportation. Jack Lemley[113], a newcomer to the project, was appointed Chief Executive.

Following the Board Meeting in Paris on 21st July, 1989, Eurotunnel announced it would be awarding the contracts for the design, manufacture and supply of the Tunnel shuttle trains at the

---

[111] He subsequently resigned the following September.

[112] Seconded from his position as Chairman of Taylor Woodrow (Scotland).

[113] Previously, Lemley had been Vice President of one of America's major engineering and construction companies, Morrison Knudsen; President and Chief Executive of the Blount Construction Group; and latterly run his own management consultancy.

end of the month. There was, however, a sting in the tail: the values of the orders would be substantially higher than previously estimated and other project costs were also likely to show increases. With the British economy passing through a period of increasing costs and rising inflation this was hardly surprising, but the statement ended by saying:

> These likely increases in costs will in the future require additional funding over and above the debt facilities already in place.

The seriousness of these financial implications was to be revealed a few days later. Costs had rocketed. The shuttle trains, estimated in November 1987 at £226 million, had spiralled to £600 million – which, together with increases in construction costs, meant that a massive £1,000 million cash injection would be needed.

It was not that Eurotunnel was short of money. Of the total loan facility of £5 billion plus £1 billion of equity, it had to date used up the equity and drawn down only £850 million of the loan facility. However, under the stringent terms of its Credit Agreement with the International Banking Syndicate, Eurotunnel must at all times show that it had sufficient finance available to fund the cost of the project to completion, and this was now clearly not the case.

Having already opened discussions with the Agent Banks, there was little likelihood of Eurotunnel being able to negotiate a further loan at the preferential rates secured at the initial 1987 financing. The alternative option, therefore, was for Eurotunnel to turn to its shareholders with a rights issue – although it had been stated that they were 'not contemplating a new share issue at this time'. Such issues are not popular with shareholders as they are usually offered at a favourable price below current market values, with the result that existing shares drop to parity with the new issue. Earlier in the year the shares, which once had peaked at £11.75, had tumbled to below £4.

The financial crisis provided a field day for the media, and initially wrecked the relationship between TML's parent companies and Eurotunnel. Battle lines were drawn and threats and counter-threats were exchanged, fuelled by rumour and speculation.

Alastair Morton fired the first broadside by laying the blame on TML's parent companies, who, accountable for the original formation of both Eurotunnel and TML, must bear full responsibility for the earlier financial estimates of the project. TML's threat to resign was countered by Eurotunnel's threat to sack TML and invite Japanese engineers to finish the job. Eurotunnel exhorted TML to cut costs; TML told Eurotunnel to look to its own backyard for economies and sack the 350-strong Project Implementation Division (PID), as it merely duplicated TML's work.

There was far too much at stake for the slanging match to continue and, having vented their respective aggressions, an uneasy truce was declared. Both sides got together to evaluate their positions, and agreed that no economies or sacrifices should be entertained which would in any way jeopardise the safety of the Tunnel or the services it would provide.

A considerable gap remained between respective completion costs: Eurotunnel's provisional estimate was £1 billion, and TML's was of the order of £1.5 billion. There were only marginal differences in respect of Target Works (tunnelling and immediately-related works) and Procurement Items (including rolling-stock), although the latter had increased significantly and orders for most such items had already been placed. The stumbling block was the likely completion costs of Lump Sum Works (the terminals and fixed equipment of the whole system).

Eurotunnel had the good grace to concede that the resolution of many of the issues had been delayed due to its insistence on the top management re-shuffle in TML earlier in the year and that therefore more time was needed to resolve these critical matters.

A meeting of a representative group of the Banking Syndicate took place in Paris on 21st/22nd September to consider Eurotunnel's financial position. This was followed by the four Agent Banks informing members of the Syndicate that Eurotunnel was no longer in compliance with certain conditions of the Credit Agreement, but it was decided that for the time being Potential Event of Default would not be notified, pending the outcome of further discussions between Eurotunnel and TML. If such Default were to be notified, Eurotunnel would have 90 days to remedy the position. There was no denying the gravity of the situation and once more the future of the project was at risk.

Subsequently, Eurotunnel and TML agreed, with the approval of the Instructing Banks, that the *Maître d'Oeuvre* should undertake an appraisal of any changes in the scope of the original Lump Sum contract and of any other relevant matters which would assist in the settlement of their differences.

In fairness to both Eurotunnel and TML, the project had been late in getting under way and there was also the problem of the shuttle train design, one of the items to suffer from early financial limitation and for which there was no precedent – in technology or experience – to call upon. This unique form of transportation had to be designed from scratch. Design, after a late start, had to be frequently changed to incorporate the many modifications demanded by Eurotunnel and the Governments' Safety Committee, which accounted in part for the cost escalation. The same could equally apply to other aspects of the project that were over-running. Estimates can only be as reliable as the information available at the time they are made.

Forethought should have anticipated the developing situation, but such was the pressure and momentum of events that any suggestion of an early pause for reflection would probably have been overruled. In those early days there was no time for calm reappraisal or for an extended planning period. Regrettably the Banking Syndicate could not afford retrospection or recognition of reasons – it needed solutions and results.

Eurotunnel's 1989 Interim Report, published on 9th October, gave a reasonably encouraging overview of progress. By 25th September nearly a quarter – 37 kilometres (23 miles) – of the total length of all three tunnels – 150 kilometres (93 miles) – had been completed.

The Report included an up-date of Traffic and Revenue studies by Eurotunnel's independent Consultants – Wilbur Smith Associates and *SETEC-Économie* – which projected a modest improvement in future revenues. They estimated (in 1989 money values) a total revenue of £578.7 million in the opening year, 1993; a revenue of £819.5 million in 2003; and a revenue of £997.4 million in 2013.

But the Report did little to dispel the murky financial cloud which hung over the project. Andrew Taylor of *The Financial Times*, who had followed the project throughout the 1980s with knowledgeable, well-balanced reporting, could not resist a cynical tailpiece to his article which appeared on 3rd October, 1989. His review of the overall current crisis concluded:

> Next week, Eurotunnel will doubtless announce an uplift in its revenue projections which may partly allay investor anxiety. But the company's record in forecasting costs hardly encourages confidence in its revenue estimates.

Although Andrew Taylor's comment was a recognisable gesture of disenchantment with the complicated flux of Eurotunnel's finances, David Danforth, Principal Associate of Wilbur Smith Associates, responded trenchantly that both his firm and *SETEC-Économie* were under contract not only to Eurotunnel but also to the Banking Syndicate, both of which had agreed to the forecasting procedures. He concluded:

> Our independent forecasts are based on an assessment of the future cross-Channel travel market, and not on the level of revenue needed to cover construction costs – we do not consider revenue needs at all! The forecasts are based on the total passenger and freight market from which we estimate Eurotunnel's potential share given our own assumptions on fares, freight rates and the services to be offered by Eurotunnel and its competitors.

There still remained the vexed question of the high-speed rail link.

There had been no further development on the possibility of a Government subvention to ease the burden of environmental safeguards. Shortly after taking office as Secretary of State for Transport, Cecil Parkinson had nothing but platitudes to offer. He said that the Government 'will do its very best' to support the Channel Tunnel project: 'It's the most important project for my Department.' But he denied that BR had told him that the high-speed line would not be commercially viable and would need a Government subsidy.

Unable to miss such an opportunity, John Prescott, Shadow Transport Secretary, responded a week later on 29th September, the eve of the launch of the Labour Party's official Transport policy. He said:

> If you want to meet the environmental needs of the South and ensure the Tunnel's benefits reach the North there has to be public money.

This could be dismissed as a point-scoring ploy, easy to make when you are in Opposition; but it was a view which claimed the growing support of back-bench MPs on both sides of the House.

The fact was that 'Green issues' were rapidly colouring the policies of both parties. The environment was becoming a central issue in the costs of future major infrastructure projects. If Governments want to attract private sector finance to such ventures, which must clearly demonstrate commercial viability, the burden of environmental safeguards must involve Government participation. The BR high-speed Tunnel link is the first example of such a project, planned before the time of change. Old values and considerations are no longer acceptable. The Government's decision on BR's current dilemma could well set the pattern for the future and caring development of a 'green and pleasant land'.

On 3rd November, 1989, British Rail announced its choice of a private sector partner for the high-speed link. The Balfour Beatty/Trafalgar House consortium, now known as Eurorail, was the successful candidate. BR further announced that the route it had proposed earlier could not be funded commercially and that it had been decided not to submit the essential Parliamentary Bill in November. It was to be postponed for a year.

One year's grace would enable BR and Eurorail to examine a more financially viable route through the suburbs of London to the terminal at King's Cross. It was the continuous tunnelling from

'Cut-and-Cover' stretch of
Tunnel between Sugar Loaf
Hill and Castle Hill.

Swanley to King's Cross which had sent the costs of the entire venture soaring beyond the point of commercial viability. With a further year's delay, however, would the new line come into operation by 1998?

In France there had been no change in SNCF's timetable. The vigorous campaign that the *Amiennois* had waged to re-route the new *TGV-Nord* line to the Tunnel terminal via Amiens had sadly failed. Even their sale of 12,000 one-metre-square plots along the preferred SNCF route had been effectively overruled, for the French Government had stepped in and granted SNCF an emergency purchase order. This order, however, did not cover land that is built upon, and in a final gesture of defiance the *Amiennois* commissioned a four-metre (13-foot) high statue in memory of Jules Verne[114], one-time city councillor, which they erected right in the middle of SNCF's route to Lille! The death of Paul Capelle, mayor of the village of Goyencourt and one of the leading protagonists in the campaign, supplies a poignant footnote to this story. He had left instructions that he was to be laid to rest on his own land – which not surprisingly was on the SNCF Paris-to-Lille route – and that his tomb should bear the provocative inscription: 'Who will dare to move me?'

On the evening of 9th November – witnessed by Cecil Parkinson and several hundred guests – the giant cutting-head of the Howden boring machine crumbled the last few feet of chalk of the eight-kilometre (five-mile) landward-drive of the pilot tunnel, breaking through at the base of Sugar Loaf Hill, Folkestone, two weeks ahead of schedule.

From this point, the pilot tunnel would continue in 'cut-and-cover' tunnelling across the narrow stretch of land which separates Sugar Loaf Hill from Castle Hill. It would deviate from its central line between the two main tunnels, passing under the south tunnel to run in parallel through Castle Hill, emerging at its base at the eastern end of the terminal site.

The purpose of this change of direction of the pilot tunnel was to facilitate the joining of the two main tunnels by the first of four crossovers which would enable trains to switch from one running tunnel to the other during maintenance or single tunnel operation.

The breakthrough at Sugar Loaf Hill coincided to the hour with

---

[114] French 19th century author and pioneer of science fiction.

a breakthrough of much greater historical significance. Nearly 600 miles away the Berlin Wall, which had divided East and West Berlin since 13th August, 1961, started to crumble. Thousands of East Berliners broke through the barrier which had constrained their freedom for so long and swarmed past the hapless border guards. It was a momentous, totally unexpected event, but what happened in the days that followed was even more so.

With astonishing rapidity, *glasnost* and *perestroika* became fact as one after the other the satellite states within the Soviet bloc began to shake free from the yoke of Communism with demands for multi-party elections – heralding a renascent democracy. Events were moving with the speed of a forest fire. One thing was vividly apparent: the mould of Communism had been broken and, whatever the outcome, Britain would be even better placed to play her role in a newly developing Europe once the Channel Tunnel was opened.

Notwithstanding these great events, the long-standing dispute between Eurotunnel and TML was exacerbated by the submission of the *Maître d'Oeuvre's* independent report to both parties on 15th December. Their estimate of £1.3 billion was very close to Eurotunnel's. The report concluded that the scope of the Lump Sum Works had not been greatly changed, and ought to be completed at a cost to Eurotunnel as provided by the contract.

An eventful year in the chequered progress of the project closed with the breakthrough, on 18th December, 1989, of the south main tunnel on the French terminal site at Coquelles. The massive 8.36-metre-diameter Mitsubishi TBM, having completed the short landward-drive of 3.2 kilometres (two miles) from the primary work site at Sangatte, would make a rapid about-turn to bore the north main tunnel from Coquelles back to Sangatte.

Finally, on 19th December, the Intergovernmental Commission and the autonomous bi-national Safety Committee accepted the principle of car drivers and passengers and coach passengers remaining in their vehicles during the shuttle train journey through the Tunnel. It had taken three years of research and demonstration to convince the authorities that this method was practical and, above all, safe.

NINETEEN NINETY

The beginning of the New Year saw little change in relations between Eurotunnel and TML. TML had agreed to an independent assessment by the *Maître d'Oeuvre*, but did not accept the findings. It was rumoured that TML would rather call a halt to construction for a cooling-off period than to continue to bear an inequitable burden. On 2nd January TML should have received £40 million from Eurotunnel, but payment had been withheld on the grounds that it was subject to current discussions on contractual issues. On 7th January TML lodged a claim for payment before the *Tribunal de Commerce de Nanterre*. This action was somewhat short-lived as the court promptly told the parties to resolve the matter between themselves, and to return only if this was not achieved.

The situation rapidly deteriorated. BBC TV's 9 o'clock news bulletin on 6th January added a touch of drama by starting a countdown:

> If Eurotunnel and TML do not reach agreement by Tuesday 9th January, the project bankers have the right to withdraw the loan facility and declare the project bankrupt.

Eurotunnel and TML met on 8th January, 1990, to find a formula the Banking Syndicate could accept. Concessions were made on both sides, but the agreement which was reached at the end of the day brought the contentious issue of Lump Sum out-turn costs no closer to settlement. The original contract price had been £1.14 billion; TML's current estimate was £1.85 billion and Eurotunnel's £1.48 billion, which the latter considered to include adequate provision for changes and contingencies. TML stood firm: it refused to make any adjustment, claiming its right under contract to submit its claim for £370 million (the difference between the two estimates) to the project's Dispute Panel and thereafter, if need be, to arbitration.

Even so, the Agent Banks passed the Agreement, with their own recommendations for consideration by the 22 Instructing Banks, at their meeting on 9th January. It was proposed that Eurotunnel's line of credit should be unfrozen, subject to conditions requiring Eurotunnel and TML to reach agreement and to exchange letters reaffirming their joint commitment to seek further reductions in direct and indirect costs.

Among the concessions that Eurotunnel had made was to reduce by some 25% the staff of its Project Implementation Division, which had become so much of a *bête noire* to TML. In December 1989, in a co-operative effort to reduce costs, it had also been agreed with TML to reduce shuttle train speeds through the Tunnel from 180 km.p.h. (100 m.p.h.) to 130 km.p.h. (80 m.p.h.). This would save money by reducing aerodynamic demands on the Tunnel's infrastructure and rolling-stock, which could in turn result in economies in design without impairing safety.

At midnight on 7th January, 1990, TML had completed and passed 50 kilometres (31.25 miles) of Tunnel, or one-third of the total to be tunnelled for the entire project. The French seaward main tunnels were going well, 9 to 14 weeks ahead of schedule, but the British seaward main tunnels were only just beginning to make real progress and were three to four months behind schedule. As tunnelling progress had been the root cause of much of the friction between Eurotunnel and TML, it was clumsy of Eurotunnel to quantify the progress of one in *weeks* and the other in *months*. In reality progress was quite encouraging and, to put Eurotunnel's schedule times in proper perspective, the overall tunnelling programme was only three weeks behind schedule. Eurotunnel seemed to have forgotten that the Tunnel was not being constructed by two separate companies, each contracted to build its own half. TML was *one* company contracted to construct the whole project, and boring operations would continue from both sides until breakthrough was completed.

There was a brief cessation of hostilities between Eurotunnel and TML, but it failed to last out the month. On 25th January the *Contract Journal* blew the lid off the recently completed Agreement by publishing 'another leaked letter', this time written by Peter Costain, Chief Executive of Costains, on behalf of the ten TML parent companies and addressed to Alastair Morton.

With the publication of the letter, whichever side had been responsible for the 'leak', no further comment was necessary. What the letter revealed was the extent of the gulf which still existed

Peter Costain, Chief Executive of Costains.

between Eurotunnel and TML – and that not only in money matters.

The adversarial role-model adopted by Alastair Morton in his dealings with TML was in some measure understandable. On behalf of Eurotunnel, he and his French Co-Chairman were directly responsible to the International Banking Syndicate and, no less important, to the shareholders, with whom a macho stewardship of their interests was possibly highly desirable. However, his comments outside the main issues of the Agreement as published in Eurotunnel's statement of 11th January were both provocative and unworthy – he had suggested that TML's British workforce lagged behind their French colleagues in efficiency, production and cost control. Eurotunnel concluded that all the present problems stemmed from TML's ten British and French parent companies, whose original cost estimates TML was now contesting.

Eurotunnel's proposed new management structure enraged TML. Alastair Morton had proposed the appointment of two additional managing directors[115] but he would be leading the team. He stepped down from his position as Co-Chairman – André Bénard remained as sole Chairman – and became Deputy Chairman and Chief Executive. Supported by five managing directors it placed him in a far more powerful position than before.

TML refused to sign and on 15th January returned to the *Tribunal de Commerce de Nanterre* to claim non-payment of £62 million, which had accrued since 2nd January. The court decided in favour of TML. With draw-down facilities from the Banks still blocked, the court's decision was academic. Eurotunnel and TML were well aware that unless they signed the new Agreement the banks would not release the funds that would enable Eurotunnel to pay TML. TML's price for its signature was the resignation of Alastair Morton.

It was a stand-off. Seeking to break the deadlock, Robin Leigh-Pemberton – Governor of the Bank of England – intervened. He called a meeting with TML and Eurotunnel on 16th February, a meeting which the five French construction companies boycotted. The meeting concluded with Eurotunnel agreeing to appoint a senior executive as a buffer between TML and Morton, who would be relieved of his day-to-day contact role. Although the British contractors agreed to this new arrangement, their French counterparts remained adamant.

That very evening, André Bénard gave a sombre TV interview. When asked whether the crisis could mean the end of the Tunnel, he agreed there was that risk: 'We shall see.' Bénard spent a day

---

[115] Alistair Fleming, ex-head of BP Exploration, was appointed MD (Construction); and Keith Bernard, late of BART (Bay Area Rapid Transit), San Francisco, was appointed as MD (Transportation).

persuading the five disenchanted French contractors to accept the compromise. The Agreement was signed by Eurotunnel and TML on 20th February. On 1st March the banks unblocked drawing facilities and Eurotunnel settled TML's outstanding February account.

John Neerhout Jnr. was Eurotunnel's newly appointed Project Chief Executive[116]. As well as his role as 'buffer' between Alastair Morton and TML, he would have full responsibility for the successful completion of the Tunnel, dealing directly with his fellow American, Jack Lemley, TML's Chief Executive.

Eurotunnel's financial situation, while temporarily on hold, would be facing major money problems later in the year. There was the need to raise sufficient funds to complete the project: figures of £2-£2.5 billion in the form of further loans and equity or quasi-equity were bruited about.

While the threat of closure once more overshadowed the project, tunnelling progress continued to improve. The British pilot tunnel TBM had broken out of the poor ground and was getting up to speed; and the pilot tunnel was closing at the rate of just under a mile (1.5 kilometres) a month, which, if maintained, forecast a November breakthrough. On 23rd April the total tunnelling distance in all three tunnels reached 75.7 kilometres (46.6 miles) – 50% of the total. Half the Channel Tunnel had been bored and lined.

This was encouraging news for Alastair Morton. Further encouragement came by way of the European Investment Bank which, in May, boosted its existing loan of £1 billion by a further £300 million.

The high rate of tunnelling progress continued and records were broken. The last week of June saw the British seaward north main tunnel TBM – following modifications – achieve 330 metres (360.9 yards) in seven days. Progress was interrupted, however, in July when the French drive on the south seaward main tunnel hit bad ground with a high volume of water flowing into the Tunnel, which slowed the progress to a rate of 20 metres (21.8 yards) per week. Shortly after this the British north seaward main tunnel TBM suffered a three-week shut-down, and in the first week after start-up achieved only 80 metres (87.5 yards). It had been stationary while waiting for the construction of the vast cavern of the crossover to be sufficiently advanced for it to pass through. The crossover cavern, 7.8 kilometres (five miles) down-tunnel from the British portal, was 163 metres (178 yards) long, 21 metres (23 yards) wide and 15 metres (16.4 yards) high; a similar crossover area would be excavated 12 kilometres (7.5 miles) down-tunnel from the French portal.

---

[116] Vice-President of the Bechtel Group, John Neerhout Jnr. had been with the Group for 24 years and headed the firm's British operations from 1979 to 1982.

The Markham/Robbins TBM 'at rest' in the huge, cathedral-like cavern of the UK crossover, 7.8 kilometres (five miles) down-tunnel from the British portal, August 1990.

By contrast, BR's high-speed link had ground to a halt. At a cost of over £3 billion the link had proved beyond the resources of any of the groups who were interested in the private financing of the project. The Ove Arup consortium had introduced a scheme which had much to commend it: environmentally sound and similar in overall distance, it would reach King's Cross via a new terminal at Stratford, but its cost was just as formidable. High interest rates dulled the gleam in the promoters' eyes and there was still no evidence that the Government would contribute towards environmental costs. BR opted for yet more studies and postponed introducing its Bill for a further twelve months until November 1991. BR still claimed that the link would be completed by 1998, but this began to look increasingly unlikely.

The Banking syndicate, on 25th October, agreed to extend to Eurotunnel an additional loan facility of £1,800 million, which, with the recent extra £300 million loan from the European Investment Bank would increase Eurotunnel's available credit from £5 billion to £7.1 billion. Eurotunnel further announced that it would be raising an additional £523 million by way of a rights issue during the period 12th November to 3rd December. The unit price would be 285p and existing shareholders would have the right to subscribe for three new units for every five shares held. The units carried new travel privileges in the form of a 50% reduction of the fare for travelling on the Tunnel's shuttle service; but, unlike the earlier offer, subscribers to the new issue could transfer their privileges to another person.

On 30th October, 1990, at 7.30 p.m., Great Britain 'ceased to be an island'. It is appropriate to recall the immortal words that symbolised Louis Blériot's cross-Channel flight in 1909, for this also was an occasion of some historic moment. For the first time in 9,000 years, Britain was physically rejoined to the Continent, if only by a probe 100 metres (109.4 yards) long and 5.08 milimetres (two inches) in diameter, 39 metres (128 feet) below the sea-bed.

The occasion was marked when, after three years of tunnelling, the British and French TBMs – having covered 22 kilometres (13.7 miles) and 16 kilometres (10 miles) respectively – came to a halt 100 metres (109.4 yards) apart. A probe driven ahead from the British side emerged on the French side to reveal that the two approaches were out of line by only 50 centimetres (20 inches)!

The British TBM was set in motion, skewed to the right of the tunnel line and entombed. The cutting head was abandoned to remain buried in the chalk and its 'train' was dismantled and taken back to the surface. The French TBM completed the drive through the last 100 metres of chalk which separated the two tunnels, but did not break through. It was withdrawn, dismantled in its entirety and also removed to the surface.

On 1st December, 1990, the official breakthrough ceremony was held. A thin diaphragm of chalk, all that now separated the two tunnels, was attacked by an air drill, wielded with much enthusiasm by 42-year-old Graham Fagg. He broke down the diaphragm and disappeared through the hole to be warmly hugged by a French miner, Philippe Cozette. No doubt overcome by the occasion, all Graham Fagg could manage to say was, 'Nice to meet you, Philippe. Are you a driver?'

Philippe Cozette's reply is not recorded, but it was explained that his understanding of English was limited. Altogether a low-key affair, at which President Mitterrand and Prime Minister Margaret Thatcher – the two principals whose joint political vision made it all possible – were not able to be present.

There can be little doubt that the breakthrough gave a timely boost to the new rights issue, which raised £566 million. With a take-up of 98%, Eurotunnel claimed it was 'a triumph of bi-national cooperation'. At the close of the year, against such a background of achievement, it could be assumed that the fortunes of the Tunnel were again buoyant and riding high. Looming over all, however, was TML's simmering resentment and its ever-growing claim against Eurotunnel, currently standing at just under £1,000 million. However, there remained a gap of some £300 million between what TML was claiming and Eurotunnel considered as being equitable. Alastair Morton firmly believed that the recently-extended loan facilities and the rights issue, together amounting to £2.6 billion, would adequately cover the total completion costs of the project and any outstanding indebtedness to TML.

NINETEEN NINETY-ONE

During the past year Alastair Morton's aggressive attitude had won him few friends. He had given – and taken – a lot of 'flak'. But when it came to money matters, even his most fervent critics would freely admit that he had the Midas touch. At the outset, project equity and loan finance had been successfully concluded with the world's money markets in total disarray, and the recent loan extension and rights issue had been nc less successful. Confidence in the project was high and Alastair Morton and André Bénard were each awarded knighthoods in the New Year's Honours list.

Relations between Eurotunnel and TML, however, showed no signs of improving. In February TML locked horns with Eurotunnel in a new confrontation. TML alleged that Eurotunnel had caused delays by 'meddling' in 1988 in the tendering process for the signalling contract, and demanded a 55-week extension of time – but not to extend the completion of the project! TML wanted the extra money this represented to accelerate the construction programme to meet the agreed completion date of 15th June, 1993. As expected, Eurotunnel refuted both the charge and the claim, maintaining that it had every right to intervene in the contract process if necessary.

Consideration of issues by a five-man Contracts Dispute Panel resulted in the most extraordinary judgement – with both sides claiming to have won. TML's claim for a 55-week extension of time was rejected; but Eurotunnel – the panel ruled – should not have interfered in the contract tendering process and should be liable to pay the cost of any delays for which it was responsible. The impact of this curious decision upon any future financial settlement remained to be seen.

Despite all the many problems, TML had always been confident in its ability to complete the project on time, but for the first time it now voiced the fear that the Tunnel's opening could be delayed.

In April this unpalatable prospect was reinforced by Eurotunnel receiving a double body blow: one from the IGC – the Governments' watchdog commission – and the other from SNCF. The IGC demanded that the width of the doors in the fire barriers between the wagons of the shuttle trains should be increased by ten centimetres (four inches). Subsequent discussions with the suppliers indicated that the necessary design modifications would mean delay

The cutting head of the Kawasaki/Robbins TBM during construction.

in the delivery of the shuttle wagons, which in turn would mean that only a limited shuttle service would be in operation for the June 1993 opening, with a full service coming into operation later in the year. SNCF, in a disturbing report, confirmed the possibility of a few months delay in the completion of its high-speed line from Calais to Paris – with trains taking some 30 minutes longer than planned.

Sir Alastair was critical of IGC's eleventh hour modification, which, together with the SNCF delay, would reduce revenues in the first year of operation by between £90 million and £100 million, but he remained optimistic that the lost revenue would soon be recovered.

The months of May and June produced the long-awaited and much overworked phrase 'light at

the end of the Tunnel'. At 10.30 a.m. on 22nd May, 1991, TML's giant Kawasaki/ Robbins TBM broke through the north main tunnel which would carry national rail and vehicle shuttle services from Britain to France. And at 11.50 a.m. on 28th June, 1991, the final breakthrough came in the south main tunnel which would carry rail and shuttle traffic from France to Britain. Both tunnels broke through ahead of schedule.

These were moments of considerable triumph for TML. Overcoming great adversities, which at times must have seemed insurmountable, it had won through within the forecast it had established six years earlier. But even this achievement was soured by a new squabble. The breakthrough of the south tunnel ahead of schedule entitled TML to a bonus of £5 million, but neither side could agree just how far ahead TML was of schedule. Eurotunnel had announced that the north tunnel breakthrough was on schedule as forecast in 1985; but the south tunnel was only three days ahead of the 1985 forecast. TML, on the other hand, firmly maintained that the south tunnel was one month ahead of the 1985 deadline and three months ahead of subsequently revised estimates. No doubt this dispute would become yet another item on the agenda.

After a chequered history spanning 189 years the Channel Tunnel was completed and lined – all 49.4 kilometres (31 miles) of it. The major problem which now faced TML, following – in Eurotunnel's own words – its 'spectacularly successful completion of the Channel Tunnel', was the challenge of the mechanical and electrical (M & E) fitting out of the tunnels and terminals. This included permanent track and the installation of 'smart' signalling, ventilation and cooling systems. It was a technical and logistical task of great magnitude, which had already fallen several months behind time.

The challenge was heightened by the fact that all was not well in the upper management echelons of TML. Its Chairman, Philippe Essig, walked out at the beginning of September. His departure came as a shock to TML on both sides of the Channel. No reason for his sudden resignation was ever revealed and rumour was rife. Pierre Parisot, deputy general manager of *Société Générale d'Entreprises* was appointed the new Chairman on 5th September. Within days of Philippe Essig's departure, Jacques Thibonnier, TML's managing director responsible for transport systems and engineering, also left and was replaced by Keith Price, a main board director of Morrison-Knudsen. TML denied that there was any connection between these two departures, or that they had anything to do with the belated M & E programme.

BR's long-running comedy of errors – the high-speed rail link – was now set to climax in the most astonishing fashion. In the Summer of the previous year, Cecil Parkinson (who was then Transport Secretary) had rejected the scheme for BR's preferred route from Folkestone to King's Cross proposed by the BR/Balfour Beatty/Trafalgar House consortium, and BR was left with a remit to come up with an alternative proposal which was more environmental and cost-friendly. In the Summer of 1991 BR submitted the new proposal – which in itself had cost £100 million – to Malcolm Rifkind, the new Transport Secretary. Having studied the report in some depth, Rifkind came to the conclusion that BR's southern route to King's Cross was the best solution. It was cheaper and the studies were far more advanced than any of the rival schemes.

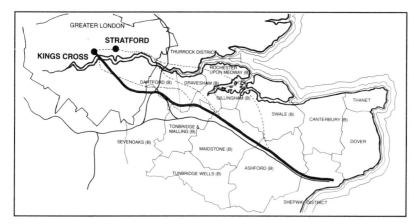

Alternative proposals for the high-speed link:

Rail link Project 1990.

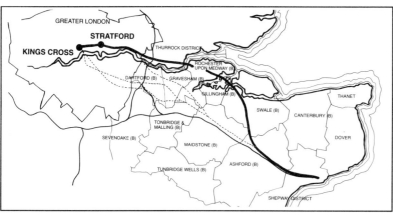

Rail Europe.

It was at this stage that behind-the-scenes politicking took over. It was suggested to Rifkind that it would pep up the rallying of the faithful if he made his statement at the Conservative Party Annual Conference at Blackpool in early October. He refused, preferring to report to the House when it resumed at the end of October. On the eve of the Party Conference, however, he found himself caught in a flurry of political in-fighting led by Michael Heseltine, Secretary for the Environment, and supported by Peter Lilley, Secretary for Trade and Industry, both fervently in favour of the 'economic urban regeneration' of the east Thames corridor. Rifkind found himself out-gunned, and on 9th October, 1991, at the Party Conference, not only did he reverse his decision in favour of the Ove Arup route which, taking a more easterly direction, would arrive at King's Cross via Stratford, but he announced that the new high-speed link would not be completed until the year 2005, when existing lines would have reached capacity!

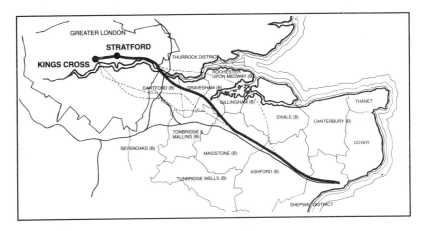

The proposed Ove Arup route.

Both BR and Eurotunnel were thunderstruck. Sir Alastair was furious. 'It is a disaster!' he said. 'The French will not know whether to laugh or cry!' He had long been of the opinion that BR's original target date of 1998 was too late by three years – but 2005 was ludicrous.

BR rightly felt betrayed by this reversal and Sir Bob Reid, when asked if he would resign over the decision, said, 'Oh, no! For heaven's sake, if you are in the middle of a pantomime you want to stay with it!' A senior BR official, who was far more matter-of-fact, maintained that the route had been chosen according to some vague political dream of economic regeneration: 'If only we had known, we would not have wasted our time and millions of pounds of taxpayers' money'.

The cost estimate of the BR route was £3.75 billion and the Ove Arup route £4.5 billion, although the latter involved less environmental disturbance. Eurorail's 1990 scheme had collapsed because of lack of Government financial support, but there was still no promise of any public sector participation, even in the form of an environmental subsidy for the latest, more costly decision. Oddly enough, each route covered the same distance of 69 miles, but the journey time of the Ove Arup route – 51 minutes – was eleven minutes longer. It could be said that this was of no great consequence; but it was in fact of prime importance to Eurotunnel who would be competing with short-haul air travel to Paris and Brussels. It had also been SNCF's experience that a saving of one minute on journey time is calculated to win 40,000 extra passengers a year.

October was a particularly bad month. Delays and disappointments had accumulated, and relations between Eurotunnel and TML erupted in a heated, public slanging match. TML threatened to stop work on the installation of the Tunnel's cooling system unless Eurotunnel increased its cash flow. Eurotunnel's immediate response was to seek a court injunction preventing TML taking any such action, a move which sent any thoughts of reconciliation plummeting. The application for the injunction was postponed on 17th October, pending further evidence. Following the news that TML's claims now exceeded £1,200 million, the British and French contracting companies which had fathered TML – angered by this new turn of events – announced that they were going to 'set the record straight with contractors, employees, shareholders and the general public' at a press conference in Paris on 23rd October. They adopted this course to prevent Eurotunnel taking action against TML who would be in breach of contract if members appeared on a public platform or talked to the media.

However, in the 24 hours before the press conference, there was a violent disagreement which split the unity of the consortium. There was a growing concern amongst the British contingent about the effect that the continued haggling between contractor and client was having on the Stock Market. Both Eurotunnel shares and the individual shares of TML's parent companies were certainly depressed. The draft speech of Jean-Paul Parayre, Chairman of Dumez, to be made at the press conference, had been circulated to the 'Ten'. The five British companies were infuriated by the aggressive tone of the draft, which they jointly agreed would only exacerbate matters further. Peter Costain demanded that copies of Parayre's draft speech should be recalled.

The 'Ten' duly appeared at the press conference, and earlier fears of an abrasive hardening of attitudes proved to be groundless. It was a public performance entirely worthy of the leaders of the ten biggest construction companies in Europe. There was no doubt as to their unity of purpose and Jean-Paul Parayre gave a no-nonsense review of TML's side of the dispute. There was plain speaking but there were no threats; if anything, there was an overriding mood of conciliation. The conference agreed that the project could be completed on time. As Peter Costain said, 'We certainly do not intend to stop work. The real question is how to find the best conditions to complete this project.'

The following day, the mood of conciliation was carried a step further by Peter Drew, Chairman of Taylor Woodrow, who wrote to Sir Alastair saying that matters were 'threatening to create a tedious and expensive impasse . . . What I'm really suggesting, Alastair, is let's cool it and quietly work together for an equitable conclusion.' Sir Alastair welcomed this approach, and in reply he said: 'I am very pleased with your wish to talk. Detailed discussions must take place.'

It was hoped that this exchange might produce a measure of badly-needed common-sense, and on 28th October the first meeting took place. As far as TML was concerned, the Paris conference had laid down four conditions which had to be satisfied if the project was to be completed successfully: (1) the project had to be clearly defined once and for all; (2) Eurotunnel should face its commitments and pay what it owed; (3) relationships between TML and Eurotunnel had to be improved; and (4) funding of the project had to be properly ensured by Eurotunnel.

Whether Eurotunnel could work within this framework, let alone reconcile its differences with TML, was uncertain. The improvement of relationships was of critical importance and could only be achieved by give and take, but there was a stark intransigence in the wording of the other three conditions.

The year ended on a more positive note for Eurotunnel. In November it signed a long-term loan facility of £200 million with the European Coal & Steel Community. Following a similar loan of £300 million from the European Investment Bank in 1990, this was the second step in the long programme of refinancing the project with long-term funds at fixed rates.

NINETEEN NINETY-TWO

Early in January the Government revived its plan to privatise the railways. Malcolm Rifkind was sorely tried by his Cabinet colleagues, who assailed him from all sides with ideas as to how it should be done. He favoured selling off BR's more profitable services, such as Inter-City and rail freight, and stubbornly resisted the various other options which were being canvassed. Following his recent experience at the Party Conference, his attitude was, perhaps, understandable. The plan was to produce a White Paper before a General Election as, come what may, the Government was irrevocably committed to face the ballot box before July. Whatever the political persuasion of the forthcoming new Parliament, it was to be hoped that there would be an early repeal of the current decision that the Tunnel's high-speed rail link would not be needed before the year 2005.

The turn of the year also saw the deferment of the phasing out of the sale of duty free goods consequent upon the introduction of the 1992 EC Harmonisation Programme. The concession would now continue until the end of the century. This was a life-line to the cross-Channel ferry operators as, without duty free sales, fares would have to be increased by at least 25%. But this reprieve did not make it any more likely that they could afford to contemplate their oft-threatened price war with the Tunnel – a threat which is surely buried in the battle-grounds of the past.

Sir Alastair, meanwhile, was preparing to open up a new battleground. His adversarial nature did not falter, neither was he deterred by the substance of his new opponent. It was revealed that he was throwing down the gauntlet to the British and French Governments' jointly appointed IGC in his pursuit of a claim 'for breach of contract in respect of its insistence on unreasonably high safety standards and its lack of regard for the commercial viability of the project'. He had been rightly angered by the IGC's last-minute demand for change in the design of the shuttle wagons. Similarly, claims for loss of revenue were being considered against BR and SNCF for their delay in the introduction of passenger and freight services.

In the original contract, December 1992 would have seen the completion of all construction work, leaving six months for TML to commission the entire system before handing over to Eurotunnel in June 1993. There was little chance of this happening, however. There were too many uncertainties – the completion of the fitting-out of the Tunnel, late delivery of sufficient shuttle trains to enable effective commissioning, or even the operation of a limited service by the opening date. The project was fast running out of time.

On 10th February Eurotunnel faced the inevitable and announced the abandonment of the 15th June opening date. It now expected a limited shuttle service to be available from December 1993, with full services coming into operation during the Summer of 1994.

March was a month of surprises. Having played such a key role in Tunnel affairs[117] in the 1970s and 1980s, John Noulton, who had joined TML as Director of Administration in 1990, announced that he was leaving to join Eurotunnel as Director of Public Affairs UK at the end of the month. The following month saw Eurotunnel's Exhibition Centre at Folkestone welcome its millionth visitor since the opening in September 1988.

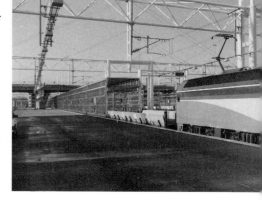

The dispute between Eurotunnel and TML rumbled on. TML had taken its claim for additional payments for

---

[117] Department of Transport.

Fixed Equipment works to the Disputes Panel and had been awarded an extra £50 million a month interim payment as from 30th April. This was a shock decision. Eurotunnel's payments for Fixed Equipment had been running at some £25 million a month and an increase of 200% was staggering. Eurotunnel immediately lodged an appeal, claiming there was no foundation for the Panel's arbitrary award. At the close of the month Eurotunnel was further notified by TML that the total value of its claims amounted to £1,217 million (at 1985 prices) with 71% relating to Fixed Equipment Lump Sum works and 15% to Terminal Lump Sum works.

On 9th April the General Election once more returned a Conservative Government with John Major as Prime Minister. The new Secretary of State for Transport was John MacGregor and the long-awaited White Paper on the privatisation of BR was published on 14th July. Not unexpectedly, it did nothing to clarify BR's position vis-à-vis the Tunnel – and was cogently summed up by Steve Wilkinson, Chairman of the Rail Development Society as 'a dog's dinner of random ideas'.

On 30th September the Arbitration Tribunal – to whom Eurotunnel had referred the Disputes Panel's decision to pay TML an additional £50 million per month – found against TML. 'Wrong in law' and 'unjustified by evidence' was the judgement. Since April Eurotunnel had overpaid TML £200 million which would be held to Eurotunnel's account by TML.

In an interim report to shareholders on 5th October, Eurotunnel reviewed the position with TML. Considerable progress had been made in refining the controversial issues with the mutually agreed intent to negotiate a binding settlement and payment agreement by the end of September. But at the last minute TML had withdrawn. Eurotunnel's peace offer of £962 million, in cash, plus £270 million in equity – which the Agent Banks considered to be generous but not to be increased – had been rejected. One of the reasons for TML's rejection was that the shares had a fluctuating value. The report went on to say that by April 1993 work would be virtually completed and commissioning underway, but until the dispute with TML had been resolved, it was difficult to consider 15th December as a firm opening date.

'Le Shuttle'

BR was able, however, to confirm that the Waterloo Passenger Terminal would be ready for the 1993 opening and that the up-grading of the existing London-to-Folkestone line was progressing satisfactorily. On the other hand, the International Passenger Station at Ashford would not be open before late 1994.

Although Eurotunnel had always been quick to react to criticism throughout the years it had been locked in its financial disagreement with TML, it had scarcely been noted for an imaginative promotion of the project. This was never more evident than in its search for a name for Tunnel services. Over a period of two years Eurotunnel had paid £500,000 to Wolff Olins – Britain's leading design agency – who had submitted hundreds of names, all of which were turned down in favour of Eurotunnel's own creation – 'Le Shuttle'! A Eurotunnel marketing executive vainly attempted to justify the choice of such a dreary name tag. 'The name', he said, 'signifies frequency, speed, flexibility and ease of use'. It certainly did not signify or reflect any of the virtues of the services to be provided, and as the vehicle-carrying trains were locked into a continuous loop they did not even resemble the characteristics of the humble shuttle. Eurotunnel had missed a great opportunity to christen its service with a name worthy of one of the greatest engineering feats of the century.

As the year came to a close, the saga of the high-speed line continued to drag on, as did negotiations between Eurotunnel and TML – with no settlement in sight.

NINETEEN NINETY-THREE

This was the year when Eurotunnel would cease to be the client of one of the century's major construction projects and become an international railway company. The man likely to head rail operations was Alain Bertrand[118].

Early in January, Union Railways – a BR subsidiary which had been formed in July 1992 with the remit to develop the high-speed link to the Tunnel – submitted proposals for a cheaper option to the Department of Transport. The following month, 22nd March, 1993, John MacGregor, Transport Secretary, gave the green light to the new link proposals at a cost of around £3 billion.

Cheaper than originally envisaged, the Government had circumvented earlier constraints with regard to public financial participation and had given a firm commitment to provide 'substantial' public funds to build the link in partnership with the private sector. The route would follow Ove Arup & Partners' original proposal and the best news of all to Eurotunnel was that it would be completed by the year 2000, five years earlier than originally envisaged by the Department of Transport. When the new line came on stream it would considerably boost Eurotunnel's revenues, for cross-Channel expresses would be able to complete the 109-kilometre (68-mile) journey from the Folkestone Terminal to London in 37 minutes. Independent studies had already indicated that the Tunnel would syphon some 80% of air passengers from the

---

[118] Born in Casablanca, Morocco, and educated in Paris, Bertrand had spent 20 years with SNCF before joining Eurotunnel as Operations Director in 1987.

'golden triangle', London/Paris/Brussels.

There was, however, an unexpected reversal in the Transport Secretary's announcement. Throughout the tortured negotiations for the high-speed rail link, King's Cross Station had been designated as the London terminal, but the Government now appeared to prefer St. Pancras! If this became the final choice it would effectively axe the £2.5 billion commercial development planned for King's Cross and surrounding area, and mean the writing off of tens of millions of pounds which had already been spent on surveys, studies and pre-planning design. This change provoked a serious re-think by all the parties concerned, but later it was announced that if this was to be the final decision BR could live with it.

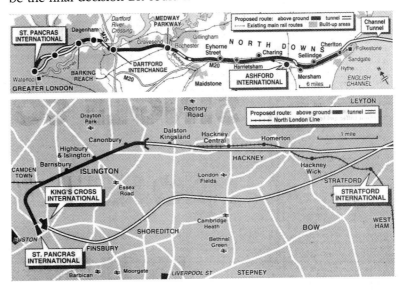

Proposed Ove Arup route to St. Pancras International.

Eurotunnel and TML continued to play the 'numbers game' of who owed what to whom, but TML remained adamant – it refused to return to the negotiating table except to discuss an improved offer.

At the beginning of April hostilities between them erupted in an explosion of threat and accusation. Eurotunnel accused TML of 'deliberately slowing down progress and employing a time blackmail.' This stemmed from Eurotunnel's claim of victory following an Industrial Tribunal ruling from Brussels that TML could no longer claim a 'global' settlement of what had now risen to nearly £1.4 billion – and it would now have to present an itemised statement of how it arrived at each claim. TML rebutted the judgement: 'They [Eurotunnel] need to deal soon if they want the Tunnel finished quickly and work could continue into next Spring of 1994.' However, TML's position was hampered by the fact that come 15th August, 1993, it would, according to contract, make its

*Aerial photograph of the
Folkestone Terminal at
Cheriton, 1993.*

first penalty payment for late completion of the project, costing
some £240 million in the first year.

Joe Dwyer, Chairman of Wimpeys and TML's Joint Chairman,
dismissed Eurotunnel's allegations as 'astounding nonsense' and
stood down from his position with TML to return in April to the less
volatile pastures of Wimpeys. On 9th May Neville Simms, Chief
Executive of Tarmac, took over his role as TML's chief spokesman
and negotiator. Further gloom descended on the project when the
Minister for Public Transport, Roger Freeman, announced on 17th
May that, although work on the International Passenger Station at
Ashford would start in October, the earliest date that the station
would open was now 1995!

Two days later, President Mitterrand underlined the shambles
of Britain's railway planning and progress as he travelled at 300
km.p.h. (186 m.p.h.) on the inaugural opening run of the 225-
kilometre (140-mile) stretch of the new *TGV-Nord* high-speed track
between Paris and Lille. The extension of the line from Lille to the
French Tunnel portal would be completed by the Autumn – seven
years ahead of anything similar between Folkestone and London!
President Mitterrand could be forgiven his gentle jibe:

> French trains will race at great pace across the plains of
> Northern France and through the Tunnel on a fast track. Then,
> on emerging in Kent, passengers will be able to day-dream at
> very low speed admiring the English landscape.

Eurostar.

May also marked the continuing exasperation of the French members of TML's construction consortium who were considering extreme measures which did not rule out a take-over bid for Eurotunnel as a final solution to the protracted dispute. Similar plans had been mooted in 1990, but this time dissatisfaction had spread to the shareholders. The Paris-based Eurotunnel Association for Action, representing some 2,000 French shareholders, threatened to boycott any future fund-raising attempt by Eurotunnel unless there was a settlement of the dispute within two months.

Against this backdrop of vexatious bickering and bitterness the Tunnel was moving slowly towards final commissioning. On 20th June the first Eurostar high-speed passenger train, built by the GEC Alsthom consortium, crept slowly but surely through the Tunnel on test, taking two hours to complete the journey. As the Tunnel's electrical system had yet to be switched on, it was hauled by a French diesel locomotive. When in operation the journey would take 30 minutes.

The financial turbulence which had raged over the project for three years was at long last quelled when, on 27th July, 1993, Sir Alastair Morton and Neville Simms met – 'without a lawyer in sight', the former commented – to sign an Agreement which would end the long-running quarrel. Behind-the-scenes meetings attended by representatives of the International Banking syndicate – who had become soured by the escalation of the conflict – had taken place under the patronage of the Bank of England over several weeks.

Under the new Agreement Eurotunnel would advance £235 million to TML who, in return, would hand over to Eurotunnel on 10th December, 1993, to complete the commissioning and final testing programme; but TML would remain responsible for the performance of the entire system. Eurotunnel stressed that the money was an advance which TML would have to justify, otherwise it would have to be repaid with interest. TML was also absolved from paying penalty payments from 15th August, unless it overran on the hand-over date of 10th December. Having settled their differences, Sir Alastair jubilantly declared that it was a 'win win solution'.

The Agreement did not, however, resolve TML's long-standing claim of some £1.4 billion which would in all probability go to Arbitration in 1994. But Neville Simms said:

> The Agreement closes the gap of whatever remains at the end. We are much more likely to find a way to resolve matters outside the Courts.

TML duly handed over the project on 10th December, 1993. The year ended on a high note with the resolution of Eurotunnel's compensation claim of some £500 million against the Joint Governments' appointed Commission (IGC) and the Safety Committee. Eurotunnel won a ten-year extension to the 55-year Concession by way of settlement. This agreement, however, did not affect the £500 million claim against BR and SNCF for their late delivery of rolling-stock and locomotives.

And so the Tunnel was finished. The inauguration by Her Majesty Queen Elizabeth II and President Mitterand was planned to take place on 6th May, 1994. Road freight and through-freight services were scheduled to open in March, with car-carrying trains to be in service by May.

Neville Simms and Alastair Morton after signing the Agreement.

Eurostar through-passenger trains would come on stream in the late summer, and coach shuttle services in September, thus completing the full range of the Tunnel's facilities.

The vision is now a reality. Though it has taken nearly 200 years, the Channel Tunnel has at last been built. It is a unique monument to Anglo/French endeavour – not just a wonderful engineering achievement, but a physical gesture of harmonous relations between two countries. From Albert Mathieu's humble plans, a massive, complex and swift transport system has come into being, with land-borne travel and trade implications between Glasgow and the whole of continental Europe.

During its long history and final construction the project has had to overcome many adversities. Probably the least of these has been technical expertise. There has never been a shortage of ideas – between 1802 and 1987 literally scores of schemes were proposed, and research suggested that many of them were workable. The greatest obstacle seems to have been Britain's determination to retain her splendid isolation. Fear of submarine invasion was to haunt British thinking from the moment Napoleon Bonaparte saw Mathieu's plans in 1802, until the moment Harold Macmillian uttered his words of reassurance in 1956. Even after that, much influential British opinion ranged somewhere between the downright obstructive and the generally apathetic. These attitudes were to keep the project at bay for another thirty years.

Public opinion has varied considerably over the two centuries. Competition with sea-ferries, the threat to the environment of Kent, and fears concerning the safety of Tunnel travel have tended to preoccupy people in recent years, and mar its potential popularity. Yet evidence suggests that there is more than enough trade to satisfy the competition; considerable effort has been made to minimise damage to the environment; and the final Tunnel incorporates within its structure and service every safety measure that can be perceived as being within the range of risk which people are prepared to accept as part of life.

The whole of life is fraught with uncertainty. The human race has gradually created a civilisation which holds at bay its more unpleasant aspects and inherent natural dangers but progress has inevitably produced further, man-made dangers.

The motor car is lethal, potentially an explosive projectile, and the aeroplane a flying bomb. Both aeroplane and ship are incompatible with their operational environment. If an aeroplane's engines fail, it falls to earth like a stone; if a ship is holed in its side, it sinks like one. If cars and aeroplanes were invented today, with the foreknowledge of their cost in terms of life and limb, would the aeroplane ever get off the ground and would the car still be preceded by a man carrying a red flag? The death toll attributed to the motorcar since it made its first appearance in 1885 is of the order of 17 million[119]. As for ships, ever since man first hollowed out a tree trunk the sea has claimed its toll. Millions of lives have been lost, and many more people injured by these three modes of transport since they were first invented.

Yet society accepts the price that has to be paid. More than three million accidents take place in British homes every year, 5,400 of them fatal[120] – which approximates the numbers killed on our roads annually. It is a sobering thought that to travel on London's Underground system is two and a half times safer than staying at home, and fifty times safer than travelling by car.

Professor Heinz Wolff, in an interesting article published in the *Daily Telegraph* Weekend Magazine of 19th November, 1988, speculated on the curious and often contradictory attitudes we take, both as individuals, and as a society, towards risk, and reached some surprising conclusions. The ability to assess risk, and to accept or reject it, is probably common to all living systems. Even play on the part of both animals and people involves deliberate exposure to quite serious risks. 'In other words', wrote Professor Wolff, 'there does not seem to be a natural human drive to reduce risk absolutely.' In fact (he points out) as the safety industry progressively removes all sorts of risks from people's lives, we spend more and more money buying them back.

Professor Wolff refers to the phenomenon known as 'risk compensation', or 'risk homeostasis' – the theory that the total amount of risk to which people are exposed remains constant, although the form of risk may change. In his review of *Risk & Freedom*, he quotes its author, Dr. John Adams of University College, London, the UK's leading exponent of this theory:

> One of the most interesting aspects of the book is a survey which Adams quotes – which compares the rate of death by accident, or violence, in 31 European, American and Australasian countries between 1900 and 1975. The surprising conclusion is that in spite of countless pieces of safety legislation, the death rates have remained constant for 75 years. The types of accident may have changed, but their number (relative to population) had not.

Paradoxically, it was Dr. Adams who, against all the evidence, so ardently supported Flexilink's claim that, in the interests of safety, drivers and passengers should be separated from their vehicles during their thirty-minute journey through the Tunnel. Professor

---

[119] Heathcote Williams '*Autogeddon*'.
[120] Department of Trade & Industry (HASS).

Wolff concludes his article with this thought-provoking summing up:

> There is little hard evidence to show that safety legislation does any good, and quite a lot of evidence to suggest that it does not. And even if it does work, a point must eventually be reached (and has probably already been reached) where the law of diminishing returns comes into operation.

The establishing of strict safety procedures in both construction and operation of the Tunnel was, and is, of prime importance. During construction the pilot tunnel performed an important function. It preceded the boring of the two main running tunnels by some five kilometres, checking ground conditions ahead of its own boring operation and also probing the ground sideways in the pathways of the main tunnels. On completion of boring operations the pilot tunnel assumed a different role, becoming the central service tunnel and providing the Tunnel complex with the essential facilities of ventilation, maintenance and all-important safety services. It is connected to the main tunnels every 375 metres (410 yards) by cross-passages; these emerge onto a platform which runs the entire length of both main tunnels. The principal ventilation of the main tunnels is caused by the piston effect of the trains which, on entering the Tunnel, start to push a column of air ahead and suck in fresh air from behind. There are piston effect relief ducts connecting the two main tunnels every 350 metres (382 yards), to allow the pressure pulse in front of the train to dissipate. This reduces aerodynamic drag, and effectively circulates the air throughout the main running tunnels. There is an additional ventilation system which would operate during emergencies, with the capability to reverse the airflow.

The service tunnel has a completely separate ventilation system, which is maintained at a slightly higher pressure and vents fresh air into the main tunnels via ducts in the cross-passages. In the event of a train being immobilised whilst in the Tunnel, the passengers would alight on to the platform and pass through the cross-passages into the clean-air haven of the service tunnel. The positive air pressure would exclude any smoke or fumes from entering the area, and passengers would be transferred to a train in the other main tunnel. There is, therefore, a continuous safety refuge alongside the trains throughout their journey through the Tunnel.

The shuttle trains are equipped with a safety curtain at the end of each shuttle wagon, capable of containing the spread of fire for a minimum period of 30 minutes. Should a fire take place in one of the wagons during the journey, shuttle train attendants – fully trained in fire-fighting – would immediately evacuate the passengers from the particular wagon and operate the safety curtains and attempt to put out the fire. The train would not stop, but continue until it was out of the Tunnel where any fire could be dealt with more effectively. Each shuttle wagon is equipped with an onboard uncoupling device which, if necessary, can be operated to isolate the affected wagon; and the locomotives at each end will pull the other wagons away in opposite directions.

Rabies has been another contentious issue. Little is known by the public about this scourge, except that it is mainly transmitted between animals and that anyone bitten by a rabid animal would share with the unfortunate creature the same high risk of an unpleasant death. However, Britain's island status, with rigorous precautions in force at ports of entry, has resisted its spread. As it is impossible to control the movement of wild animals across land frontiers, rabies is very prevalent on the Continent; and it has been claimed that the Tunnel will of itself make Britain vulnerable to the disease. In fact, the Tunnel will make Britain no more vulnerable to rabies than, say, a ferry vessel carrying road vehicles, for it embodies special precautions as recommended by the State Veterinary Service. The Terminal area is surrounded by a close-meshed security fence, bedded deep into the ground below the reach of burrowing animals, and armed with detection devices to warn of the approach of other creatures. Inside the Terminal area there are dog patrols, infra-red TV surveillance, and electrified grid pits to prevent animals gaining access to the Tunnel itself. Nothing attracts animals more than food, and the interior of the Tunnels is kept scrupulously clean. The windows of all Tunnel trains are sealed to prevent food being thrown out onto the tracks.

Eurotunnel's Veterinary Adviser, A. J. Crowley, former Veterinary Head of the Rabies Section, State Veterinary Service, Ministry of Agriculture, Fisheries and Food has stated:

> The measures proposed to protect the tunnel against entry to it, and the passage through it, of wild or stray rabies-susceptible animals will, in my opinion, be adequate and effective and will ensure that the tunnel will not pose any increased risk of the introduction of rabies into the UK over and above any that exist at present.

In the future, ultrasonics could well make a useful contribution. There are already devices on the market about the size of a cigarette pack which it is claimed will effectively keep a house free from mice, rats and even cockroaches. Furthermore, the Pasteur Institute in Brussels is actively getting to the root of the problem. It has already developed a new anti-rabies vaccine and is conducting a number of field trials using vaccine-impregnated food pellets which are scattered around the woods and countryside. If this vaccine is successful *la rage* – as rabies is more appropriately known in France – may well be eliminated altogether. However, until a successful antidote is discovered, it must be recognised that the worst offenders, and the most likely source of re-infecting Britain's animals with the rabies virus, are those so-called 'pet lovers' who smuggle their pets ashore at any one of the numerous sheltered bays around the coastline.

Two other issues may be seen to have overshadowed the sense of achievement. The first has been the highly-publicised three-year dispute between Eurotunnel and TML concerning money and deadlines. An eminent civil engineer once declared, 'Any major project which is completed before time and under budget must have been overestimated at the outset!'

The cause of the late opening was twofold: according to the original contract, the construction of the Tunnel and its terminals was to be completed by December 1992, followed by a six-month commissioning period, with an opening date of 15th June, 1993. The completion of construction was, however, four months late (April 1993) – not exactly an indictable offence over a six-year period of construction; but the real problem – due, it was said, to late changes in design – was the delayed delivery of the special vehicle-carrying rolling-stock and locomotives, and of the Eurostar passenger trains. All these should have been delivered to coincide with the original commissioning period in the first six months of 1993. Hence, although the Tunnel itself was completed, services were delayed for nearly a year with only sufficient dedicated rolling-stock to see the start-up of a phased service in March 1994.

It is doubtful that the Channel Tunnel will ever be recognised as a blue-print which might inspire the private financing of major projects in the future but it does spell out the financial pitfalls that should be avoided, and underlines the necessity for meticulous pre-planning. However, to have interrupted the momentum of the project at the outset with a year's calm appraisal would undoubtedly have led to yet another abandonment of the Tunnel.

The other issue has been the continual shilly-shallying over the building of Britain's high-speed rail link; and it has to be said that the full benefits of the Tunnel, both in terms of personal and business travel and in particular of railway freight transport, cannot be realised until that link is established. To a large extent Britain's roads will remain clogged with freight lorries; and the planned journey times of London-to-Paris 2 hours, 27 minutes, and London-to-Brussels 2 hours, 7 minutes, will not be achievable until the link is built.

Despite its critics, and the shortcomings of its enforced phasing of services, the Channel Tunnel is a wonderful achievement, a fitting tribute to such men of vision as Albert Mathieu, Thomé de Gamond, Sir Edward Watkin, Albert-Henri Sartiaux, and many others. It is a national asset – an economic umbilical cord which, in terms of convenience and the rapid transit of people and goods, will add a new dimension to international travel. A little late in its arrival, it emerges at a time when Britain is forging ever-stronger links with the continent.

The Tunnel will continue to have its critics, and Kent will continue to harbour pockets of opposition, but the question on the lips of millions will soon be: 'Whatever did we do before we had a Channel Tunnel?'

VARIATIONS ON AN OLD SCHEME

For nearly two centuries the challenge of linking Britain physically with the Continent produced a proliferation of ideas and schemes. These proposals have at one time or another been described as original, impractical, ingenious, eccentric, bizarre and crackpot. The later history of the project proved equally fertile ground. It would be inappropriate not to mention a few of these ideas – some 'variations on an old scheme', others more recent.

HOREAU'S TUBE

In 1851 the *Illustrated London News* published an account of the fascinating scheme of Hector Horeau of Paris:

> M. Horeau's project consists in crossing the English Channel . . . by means of a tube made of strong plate iron, or cast iron, lined and prepared for that purpose: and which, placed at the bottom of the sea should, beside the path for the surveyors, contain the two lines for the trains which would run within it. The slope given to the submarine railway would admit of a motion sufficiently powerful to enable the carriages to cross the Channel without a steam-engine. The greatest depth of the sea at the middle of the Channel will admit of the construction of inclined planes, by means of which the train would be enabled to reach a point where a stationary engine, or atmospheric pressure, might be employed in propelling the train to the level of the land railways of France and England. These tunnels beneath the sea would not prevent navigation . . . According to an estimate made, the costs might amount to about £87,400,000.[121]

The accompanying drawings show a whole string of bell tent-shaped, neo-Gothic pavilions serving as lighthouses as well as ventilation shafts, which represent a veritable nightmare for the Channel sea captain. Under the sea, too, there would be formidable dangers, particularly for Horeau's trains; and for these reasons the scheme was soon dismissed.

THE FLOATING ANGLO-GALLIC SUBMARINE RAILWAY

In 1855 the British engineer James Wylson wrote to the *Illustrated London News* in order to dispute the efficiency of trains operated by gravitational force. He then outlined his own scheme for a tunnel floating mid-way between the sea bed and the surface, held in position by ties to the sea bed and to buoys above. There would only be one set of rail tracks and the trains, which would operate alternately from each shore line, would be powered by a 'push-pull' locomotive.

James Wylson's cost estimate was £15 million, without the rolling-stock; but his floating Anglo-Gallic Submarine Railway, as he called it, with its electric locomotive, reading lights and glass walls was certainly a splendid vision of the future.

---

[121] The writer may have confused pounds sterling with French francs, as the cost would have been astronomical for the mid-19th century.

THE ICE TUNNEL

This was the most original and the most improbable scheme ever to be proposed, but anyone familiar with the relatively unknown World War II exploits of Geoffrey Nathaniel Pyke[122] would have observed that bizarre as it might have appeared, it had a background of scientific substance.

When Lord Louis Mountbatten became Chief of Combined Operations, he formed a 'think tank' of free-ranging, unconventional minds – unusual people with unusual talents – to pursue his policy of positive warfare. In this atmosphere Geoffrey Pyke found an outlet for his outstanding talents as one of Mountbatten's three principal scientific advisers, together with Solly Zuckerman and Professor Burnal, the London University physicist. It was Burnal who maintained that Pyke was 'one of the greatest geniuses of his time'.

Pyke discovered that by freezing a mix of sawdust and water, the resultant material was capable of a crush-resistance of more than 3,000 lbs per square inch. A column of the material – subsequently named Pykrete – one inch in diameter, was capable of supporting, with ease, the weight of a medium-sized car.

The application of Pykrete would have revolutionised the construction of ships, aircraft-carriers and floating aircraft landing strips. Warships would have the advantage of an extremely low profile as, like an iceberg, seven-eighths of their bulk would be below the water-line and they would be virtually unsinkable. Refrigerating plants linked to a network of small-bore pipes contained in the frozen hulls of ships would re-freeze any damage sustained below the water-line.

In fact Allied War Chiefs put into effect 'Operation Habbakuk', the construction of a 2,000-foot-long aircraft carrier with a displacement of over two million tons. Events subsequently overtook the Operation – but not before construction was far advanced.

It was against this background that in 1966 an Israeli engineer named Tilman approached the Channel Tunnel Study Group with his patented scheme for building the Tunnel of ice! He proposed a circular framework of small-bore pipes laid on the sea bed, through which would be pumped ammonia gas. The water around the pipes would freeze, thus forming a tunnel of ice which would be suitably insulated and lined. The pumping stations to maintain the flow of gas through the pipes would be situated at either end of the tunnel. They would be gigantic structures – 6,000 feet square and 30,000 feet high – occupying over a square mile of ground and towering nearly six miles high. The towers would consist of a mass of linked vertical tubes. Gas would be pumped to the top of these tubes where, passing through a condenser, it would return on its downward path in liquid form; in the process it would operate a series of hydro-turbines to be pumped into the pipe network under the sea, thus maintaining the refrigeration of the tunnel walls. The hydro-turbines would produce a valuable and free by-product in the form of millions of kilowatts of electricity.

Tilman also claimed that he had overcome the fears of the Tunnel's possible use by an invasion force from Europe. 'In the event of war', he said, 'you just turn off the gas and within fourteen days – no tunnel!'

The principle was simple enough, but even if feasible, at a cost of £200 million the timing was inopportune. The practical and political progress of the Tunnel was too far advanced for such a science fiction method of construction to be contemplated.

---

[122] *Pyke, the Unknown Genius* by David Lampe.

THE FLOATING ROAD TUNNEL

This was to be a submarine tube 50 feet in diameter, constructed of one-inch-thick steel plate with ten-inch-deep corrugations to maintain rigidity. It would be supported by its own buoyancy some 45 feet below low-water level and anchored in position by steel cable ties fixed to concrete piles at intervals of 30 feet. To reduce the overall specific gravity and achieve the requisite degree of buoyancy, considerable quantities of ballast would be needed which would provide a convenient bed on which to lay a 40-foot-wide, four-lane motorway.

THE SQUARE TUNNEL

There were several proposals for square tunnels, but the method put forward by Christiani & Neilsen, the Danish-based company of international repute, is worthy of record. This company submitted to the Channel Tunnel Study Group plans for a tunnel – oblong in cross-section, with two internal divisions providing three square-shaped tunnels – at half the cost of bored and immersed tube tunnels under consideration by the Group. The square tunnel concept also made the best use of available space – road and rail vehicles being rectangular in profile – and as a result aroused more than passing interest.

There were some unique features embodied in the plans. Huge 500-foot-long pre-stressed concrete oblong box sections would be prefabricated on both sides of the Channel, each with a rail-track built on to its upper surface to facilitate laying the sections. After positioning the first section, the second, supported on bogeys, would travel down the rail-tracks on the topside of the first. At the end of the track it would be lifted clear by an overhead, self-propelled gantry; lowered into a prepared trench; and joined to the previous section.

One-third of the tunnel would be sunk in a prepared trench, with two-thirds standing proud of the sea bed. At this juncture the attraction of the cost benefit began to fade. The Department of Transport's prescribed safety requirements would not permit any part of the Tunnel to project above the sea-bed. It would have to be completely buried in a trench under a minimum thickness of five metres (16 feet) of rock-fill throughout the entire length.

BRIDGE-ISLAND-BRIDGE

Undeterred by the cumulative problems associated with 'bridging the gap', technological uncertainties, navigational hazards, financial viability, and international agreement, 'bridge buffs' continued to promote their ideas with vigour and enthusiasm.

An island in mid-Channel, linked by tunnels to Britain and France, was first proposed by Mathieu in the nineteenth century. The combination of an island linked by bridges, however, was the vision of Willem Frischmann – senior partner of Pell Frischmann, a London-based company of engineering consultants – who claimed that theirs was no idle, futuristic 'doodle'; its application was to the present day.

The bridge[123], 230 feet above the sea, was basically a steel tube tunnel on stilts, changing to suspension support for the centre spans; but the most striking feature was the 30-square-mile island which would be reclaimed and built on the Varne and Colbart Banks. On either side of the island the bridge would have a clear three-mile span suspended from 1,500-foot-high towers. It was claimed that these would present no hazard to shipping and they would clearly define the main 'up' and 'down' deep-

---

[123] A modified version of this bridge – without the island – became the Eurobridge scheme of the 1980s.

water lanes. Stretching back to the coasts, the bridge spans would be reduced to a quarter of a mile in length, supported on concrete piers.

The enclosed bridge deck, made from welded steel plate, and clad with corrosion-proof steel sheet, would look like a huge oval tube, with circular opens and windows along the sides, and with tinted glass in the roof to reduce glare and solar heat. The shape would provide maximum rigidity; and the openings in the side, in addition to reducing aerodynamic forces, would furnish both view and ventilation, and would ensure an all-weather, 365-day operation.

Inside the oval tube there would be two road-traffic decks, each with two three-lane carriage ways, and railway tracks would flank the upper-deck carriage ways.

MID-CHANNEL CONSTRUCTION

In May 1983, the Davidson brothers who, with their company Technical Studies, Inc., NY, had figured so prominently in the early days of the Channel Tunnel Study Group, returned to the scene with a proposal for an accelerated tunnelling method. It had been invented by Frank Davidson, President of TSI and developed by a subsidiary company, Mid-Channel Access Corporation. It was an ingenious method of providing six additional working faces which would reduce overall tunnelling operations and costs by two years. The resulting saving of interest on loan capital and the Tunnel coming on line in a revenue earning mode – two years ahead of schedule – were both attractive.

Their innovation consisted of a massive concrete or steel caisson sunk in mid-Channel and cut into the sea-bed to the level of the tunnel line, with a minimum top clearance of 20 metres below low-water level, so as not to interfere with surface shipping traffic. Access to the caisson would be a series of air-locks through which the boring machines would be lowered to their work face – boring towards the coast in both directions. Spoil from the boring machines would be piped along the sea-bed to the nearer coastline for disposal in the form of slurry.

This was not an entirely original concept, as de Gamond, in the nineteenth century, had conceived the idea of providing two additional work faces for tunnelling by sinking a shaft on the Varne Bank. The proposal was examined in some detail by CTD 81 who came to the conclusion that costs of US$100 million had been underestimated, even when the additional cost of a further six boring machines had been discounted.

TRANSPORT
RESEARCH

French Railways (SNCF) decided in 1969 on a thorough structural re-organisation of its outdated system. Much of the planning and design work for a faster future owed its impetus to Roger Hutter, Joint Director-General of SNCF. Gas-turbine trains had been developed on medium traffic lines offering improved speeds and amenity standards.

It was during this period of planning and experiment that the potential of tracked hovercraft was investigated. Professor Colin Buchanan, in a technical paper that he produced in 1969, advocated that the most promising application of tracked hovercraft would be its introduction as a form of high-speed transport between main centres of population.

Although France was the first country to consider the practical application of linking the two airports of Orly and Roissy by the *aerotrain*, it was Britain which was to contribute the ideal power system for the hovertrain. Professor Eric R. Laithwaite of London's Imperial College developed the linear induction motor which has no moving parts, noise or exhaust gases, and is capable of speeds of up to 300 m.p.h.

In the Summer of 1971, Britain's prototype hovertrain began its first test run, but it was abandoned by the Department of the Environment on 14th February, 1973, on the grounds that there appeared to be no practical application for hovertrains in the foreseeable future.

In 1971, the Council of Europe issued a report which envisaged a tracked hovercraft line – a hovertrain – from Brussels via Luxembourg, Strasbourg and Western Switzerland to Geneva: a journey of 500 miles which could be accomplished by such a train in two hours.

A more recent technique, 'peristaltic propulsion' – which took its name from the human body's muscular, wavelike contractions of the alimentary canal – was developed by John F. Spillman, Senior Lecturer at the College of Aeronautics at Cranfield, Bedfordshire. Based on a reverse of the hovercraft principle, a vehicle in the shape of a platform, or a purpose-designed container with no working parts, is supported and propelled on a cushion of air. The first thrust stage produces an air cushion beneath the vehicle, by compressed air being directed upwards from angled nozzles situated in the track. The second thrust stage is achieved by the compressed air escaping from the vehicle itself through rearward-facing nozzles.

BR's contribution to the high-speed development of rail transport was the Advanced Passenger Train (APT-E). The prototype of the APT-E began its experimental runs on the Melton Mowbray test track in 1971, with a view to entering service in the 1980s. There were two versions: one powered by gas-turbines for operation in Britain; and the other electrically-powered for 25 k.v. overhead transmission, to be used in the Tunnel.

The APT had a top speed of over 160 m.p.h. and the articulation bogeys would allow the axles to take up different attitudes, enabling high-speed to be maintained on curves.

The APT – BR's flagship of the future – was a unique answer to the successful French TGV. The 'tilting' mechanism enabled the APT to operate on existing tracks, whereas the TGV required the additional expense of new tracks with minimum curves. Sadly it never fulfilled its great promise.

On its inaugural run, carrying representatives of the world's transport press, it broke down more than once, and eventually limped back to base. Despite further endeavours to iron out the erratic behaviour of the 'tilting' mechanism, the APT never entered service and the project was abandoned.

However, in the 1980s BR believed that it could maximise the opportunities for the Continental passenger and freight services which the Tunnel offered – at least for the remainder of the century – with the up-grading of existing rail tracks between London and Folkestone. This 'no new high-speed track' alternative led to its commitment with SNCF for the joint development and construction of a new generation of high-speed, multi-voltage locomotives and rolling-stock, built to BR's loading gauge and dedicated to Tunnel operation. The new locomotive would have the capacity to operate from both third rail and overhead power transmission and would be adaptable to other changes of voltage on the Continent. Some fifty sets of these new trains, capable of speeds of 280 km.p.h. (180 m.p.h.), would come into service in 1994 when the Tunnel opened and would be operated by British, French and Belgian railways. Presumably they would continue in service irrespective of BR's change of policy in 1989 to build a new high-speed track.

The space-age train of the twenty-first century, if research and development continue to demonstrate promise, will be 'powered' by electromagnetic levitation – (Maglev) – which bears some relation to Professor Laithwaite's development of the linear induction motor. The Maglev train concept, 'Transrapid', is being developed by Transrapid International, founded by a consortium of leading German engineering companies and based at the 13.5 kilometre test track facility at Elmsland, Lower Saxony. Originally designed to operate on a specially designed elevated guideway, similar to a monorail, the capability of 'Transrapid' to adapt to branching tracks was developed.

This led Transrapid International to contact the Channel Tunnel Study Group in 1984. The internal diameter of the Tunnel at that time was 7.3 metres (24 feet) which, much to the Germans' disappointment, was some 15.2 centimetres (six inches) less than would be needed to operate 'Transrapid' through the Tunnel. However, since that time the internal diameter of the main running tunnels has been increased to 7.6 metres (25 feet) which presumably would be adequate to accommodate 'Transrapid'.

Environmentally, 'Transrapid' – capable of speeds of 400 km.p.h. (250 m.p.h.) – has much to offer: no exhaust gases or fumes and, more particularly, dramatically reduced noise pollution. Electromagnetic suspension, with no rail contact in travel, produces extremely low noise levels, with an energy consumption comparable with a car. At 300 km.p.h. it generates no more noise than a car at 25 metres distance. Up to distances of 800 kilometres (500 miles) it will narrow the gap between current high-speed trains and short haul air journeys. One readily recognisable route under consideration is from Los Angeles to Las Vegas, a distance of 400 kilometres.

Both technically and economically, wheel-on-track locomotives are reaching their optimum with speeds around 180 m.p.h. Metal on metal friction not only sets the limits to the speed that can be obtained, but entails high wear-and-tear operating costs. While the costs of track for the Maglev system are about the same as for conventional rail tracks, operational costs are considerably lower with its economical use of energy and its friction-free suspension, guidance and propulsion.

The next generation of high-speed trains after the legendary French TGV and the Japanese Shinkansen – which revolutionised rail travel in the latter part of the twentieth century – could be the 'Transrapid'. Perhaps it is not too fanciful to speculate that this might see light at the end of the tunnel in the twenty-first century.

EXTRACTS FROM THE
ECONOMIC &
FINANCIAL STUDIES
PUBLISHED BY THE
BRITISH CHANNEL
COMPANY

9TH JUNE 1973

The Cheriton Terminal is closer to the Tunnel portal than was earlier planned, and the design and operation has been refined providing a peak-hour service at a two and half minute headway. This capacity is approximately equal to that of the dual-carriage, three-lane M20 motorway which is to be built, irrespective of there being a Tunnel, from London to Folkestone.

The actual travelling time between terminals would be approximately thirty-five minutes and three types of road vehicle shuttle trains would operate round the Tunnel circuit.

TRAFFIC AND REVENUE

On the basis of the final studies completed in 1973 by the independent consultants Cooper & Lybrand Associates Ltd. and *SETEC-Économie*, the revenue and operating estimates for 1981 (the first full year of operation) and 1990 would be as follows:

	1981 (£m)	1990 (£m)
Gross revenue	122	286
Operating cost	17	34
Operating profit	95	252
Debt service	69	89
Net receipts	26	163

CAPITAL COST

The out-turn cost of the project in 1980 – the year of completion – was estimated to be £846 million:

1. Construction costs at January 1973 prices.
2. Provision for interest, inflation and other financing costs during construction.

    Total cost to be shared equally between the British and French Channel Tunnel Companies.

FINANCE

A minimum of 10% of the cost would be risk capital provided by the British and French Channel Tunnel Companies, the remaining finance, without recourse to public funds, would be raised on the international money markets by the two Companies as and when required during construction in the form of bond issues which would carry a joint Government guarantee.

TRAFFIC FORECAST

Forecast Tunnel traffic for the first full year of operation and for 1990:

	1981	1990
Passengers without cars:	9 million approx.	13 million approx.
Passengers with cars:	6 million approx.	11 million approx.
Freight:	4½ million tons approx.	8 million tons approx.

The Tunnel traffic forecast was based on the conservative assumption that the economic growth (GNP) of the UK would only be 2.8% annually and that of France 4%.

TUNNEL TOLLS

The Tunnel revenue forecasts were based on a toll charge 18.5% lower than prevailing charges on the short sea-ferry routes in 1973.

Directors and Principal Officers of the Channel Tunnel Group and of Eurotunnel

CHANNEL TUNNEL GROUP 1984

Directors

Christopher Chetwood (later knighted) Chairman, George Wimpey; Denis Child Director and Deputy Group Chief Executive, National Westminster Bank; Sir Frank Gibb Chairman, Taylor Woodrow; Donald Holland Chairman, Balfour Beatty; Alan Osborne Managing Director, Tarmac; Christopher Tyrell Wyatt Chairman, Costain.

Executive Committee

Tony Gueterbock Wimpey, Martin Hemingway Tarmac, David Hoggard Balfour Beatty, Colin Kirkland Mott Hay & Anderson, Ted Page Taylor Woodrow, John Sargent Costain, Colin Stannard National Westminster Bank, and John Taberner Costain.

EUROTUNNEL 1986

The joint board of directors of Eurotunnel PLC and Eurotunnel SA – the Anglo/French partnership responsible for the private financing, construction and operation of the Channel Tunnel – was announced on 25th September, 1986.

Company Secretary
David Wilson

Treasurer
Peter Ratzer

The following is a list of Co-Chairman 1986-87:

Co-Chairmen	Appointed
Lord Pennock of Norton Director, Morgan Grenfell PLC	29th May, 1986 (resigned as Co-Chairman 20th February, 1987, but remained on Board)
André Bénard Member Supervisory Board Dutch Petroleum Co.	25th September, 1986
Denis Child Director, Deputy Chief Executive National Westminster Bank PLC	29th May, 1986

Alexandre Dumont   25th September, 1986
President *Belgamanche* SA

Jean Fontourcy   25th September, 1986
Deputy Chief Executive   (resigned as Co-Chairman
Caisse Nationale de Crédit   20th February, 1987)
Agricole

Sir Alistair Frame   25th September, 1986
Chairman Rio Tinto Zinc Corporation PLC

Sir Nicholas Henderson   25th September, 1986
Former British Ambassador,   (resigned as Co-Chairman
USA Ex-Chairman of   28th July, 1988)
Channel Tunnel Group

Jean-Paul Parayre   1st September, 1986
Chief Executive *Dumez SA*
Ex-President *France-Manche SA*

Patrick Ponsolle   25th September, 1986
Deputy Chief Executive   (resigned as Co-Chairman
Compagnie Financière de Suez   20th February, 1987)

Bernard Thiolin   1st September, 1986
Chief Executive
Crédit Lyonnais

(Now follows a list of subsequent Board appointments prior to the launch of
Equity 3)

Sir Nigel Broackes   17th November, 1986
Chairman Trafalgar House PLC   (resigned 2nd February, 1987)

Jean-Loup Dherse   18th December, 1986
Chief Executive Eurotunnel   (resigned 27th January, 1988)

Michael Julien   18th December, 1986
Deputy Chief Executive   (resigned 27th January, 1988)
Eurotunnel

Alastair Morton   20th February, 1987
Chairman Guinness Peat PLC
Co-Chairman Eurotunnel PLC

Bernard Auberger   20th February, 1987
Chief Executive *Caisse*   (resigned 29th November, 1988)
Nationale du Crédit Agricole

Bernard de la Genière   20th February, 1987
Chairman & Chief Executive   (resigned 28th July, 1988)
Compagnie Financière de Suez

Robert Lion   20th February, 1987
Chief Executive
Caisse des Dépôts et Consignations

Sir Kit McMahon   20th February, 1987
Chief Executive & Chairman
designate Midland Bank PLC

Robert Malpas   5th April, 1987
Managing Director
British Petroleum PLC

Pierre Durand-Rival   14th October, 1987
Deputy Chief Executive Eurotunnel

Arranging and Syndicated Banks

Project finance credit facilities in loans and letters of credit for a total of £5 billion (£2,600 million, FF21,000 million and US$450 million) for construction of the Channel Tunnel were simultaneously signed in London and Paris between some 200 international banks and Eurotunnel Finance Limited and Eurotunnel SA on 4th November, 1987. The total of £5 billion includes the £1,000 million credit arrangements with the European Investment Bank.

ARRANGING BANKS

Credit Lyonnais, National Westminster Bank PLC, Banque Nationale de Paris, Midland Bank plc, Banque Indosuez

UNDERWRITING BANKS

Banque Indosuez Group, Banque Nationale de Paris, Credit Lyonnais, Midland Bank plc, National Westminster Bank PLC, Amsterdam-Rotterdam Bank N.V., Arab Banking Corporation (ABC), Banca Commerciale Italiana, The Bank of Tokyo, Ltd., Barclays Bank PLC Group, Bayerische Vereinsbank A.G., Credit Agricole, Citibank, N.A., Commerzbank A.G., Deutsche Bank A.G., Dresdner Bank A.G., The Industrial Bank of Japan, Limited, Lloyds Bank Plc, The Long-Term Credit Bank of Japan, Ltd., The Sanwa Bank, Limited/Sanwa International Limited, Security Pacific National Bank, The Tokai Bank, Limited, Union Bank of Switzerland, Credit Suisse, The Dai-Ichi Kangyo Bank, Limited, The Daiwa Bank, Limited, The Fuji Bank, Limited, Generale Bank S.A./N.V., The Mitsubishi Bank, Limited, The Mitsui Bank, Limited[124], Banque Arabe et Internationale d'Investissement, Den norske Creditbank Group, Hessische Landesbank Girozentrale[125], Kredietbank International Group, The Mitsubishi Trust and Banking Corporation, The National Bank of Kuwait S.A.K., The Saitama Bank, Ltd[126]., The Taiyo Kobe Bank, Limited[127], Westdeutsche Landesbank Girozentrale, The Bank of Nova Scotia, Canadian Imperial Bank of Commerce (International) S.A., Credit National, NMB Bank, The Nippon Credit Bank, Ltd., Standard Chartered Bank, The Sumitomo Bank, Limited, The Yasuda Trust and Banking Company, Limited, Banque Internationale a Luxembourg S.A., BIAO-Afribank, AL UBAF Banking Group

SENIOR MANAGERS

Kreditanstalt für Wiederaufbau, Moscow Narodny Bank, Limited, DG BANK INTERNATIONAL S.A., S.N.C.I.-N.M.K.N., Banque Federative du Credit Mutuel, Banque Française du Commerce Exterieur, Credit du Nord, Credit Industriel et Commercial de Paris, Banque Commerciale pour L'Europe du Nord (EUROBANK), Union Bank of Norway, Arab Bank, Limited, ASLK-CGER Bank, Banco di Napoli, Bank of China (London and Paris), Banque de l'Union Europeenne, Consorzio di Credito per le Opere Pubbliche – CREDIOP, Credit Communal de Belgique S.A./Gemeentekrediet van Belgie N.V., EFIBANCO S.p.A., Girozentrale und Bank der österreichischen

---

[124] Name has changed to Sakura Bank Ltd.
[125] Name has changed to Landesbank Hessen-Thueringen Girozentrale.
[126] Name has changed to The Kyowa Bank Ltd.
[127] Name has changed to Sakura Bank Ltd.

Sparkassen AG, The Hokkaido Takushoku Bank, Limited, The Kyowa Bank, Ltd., The Mitsui Trust and Banking Co., Ltd., Groupe Societe Generale, Swiss Bank Corporation, TSB Group

MANAGERS

Banca Popolare di Milano, Banco de Bilbao, Banque de la Societe Financiere Europeenne, Bayerische Landesbank Girozentrale, The Chuo Trust and Banking Company, Limited, Creditanstalt Bankverein, Genossenschaftliche Zentralbank AG, Kansallis Banking Group, The Royal Bank of Scotland plc, Union Bank of Finland Ltd.

SENIOR CO-MANAGERS

BACOB Savings Bank S.C., Banco Hispano Americano Group, Banco di Sicilia Group, Bank of Scotland, Banque Regionale d'Escompte et de Depots (BRED), Berliner Bank A.G., Caisse Centrale des Banques Populaires, Cassa di Risparmio di Torino, Götabanken, The Hyakujushi Bank, Ltd., ICCR-Istituto di Credito delle Casse di Risparmio Italiane, Osterreichische Länderbank, Royal Trust Bank

CO-MANAGERS

Alahi Bank of Kuwait KSC, Cie BTP-Finance/Banque du Batiment et des Travaux Publics, Skandinaviska Enskilda Banken, The Ashikaga Bank, Ltd., Banco di Santo Spirito, Bank für Gemeinwirtschaft AG, The Bank of East Asia Limited, The Bank of Yokohama, Ltd., Banque Cantonale Vaudoise, Banque Demachy et Associes Paris, Banque Generale du Luxembourg S.A., Banque de Neuflize, Schlumberger, Mallet, Caisse d'Epargne de l'Etat du Grand-Duche de Luxembourg/Banque de l'Etat Caisse d'Epargne Geneve, The Chiba Bank, Ltd., Citic Industrial Bank, The Commonwealth Bank of Australia, Limited, Credit Chimique, Den Danske Bank, Deutsche Girozentrale-Deutsche Kommunalbank, DBS Bank, Electro Banque, The Hachijuni Bank, Ltd., The Hokkaido Bank, Ltd., The Hokuriku Bank, Ltd., FRAB Bank International, Hamburgische Landesbank Girozentrale, Investors in Industry PLC, Kuwaiti-French Bank, Riyad Bank, The Rural & Industries Bank of Western Australia, Sparkassen SDS, Swiss Cantobank (International), Swiss Volksbank, The Toyo Trust and Banking Company, Limited

PARTICIPANTS

Banca del Gottardo, Banco Exterior Group, Bahrain Middle East Bank (E.C.), Bank Leu Ltd., Die Erste Osterreichische Spar-Casse-Bank, Banco Arabe Espanol S.A., The Bank of Fukuoka, Ltd., The Bank of Hiroshima, Ltd., B. Metzler, seel, Sohn & Co KGaA, The Nippon Trust Bank, Limited, Al Saudi Banque, Zentralsparkasse und Kommerzialbank, Allied Irish Banks plc, Baden Wurtemmbergische Bank AG, Badische Kommunale Landesbank Girozentrale, Banca Credito Agrario Bresciano, Bank für Handel und Effekten, The Bank of Kyoto, Ltd., Credit Cooperatif, Banque Hervet, Banque Industrielle et Mobiliere Privee, Banque Petrofigaz, Banque Sudameris France, Banque Worms, Bergen Bank A.S., Cassa di Risparmio di Genova E Imperia, Cassa di Risparmio di Verona Vicenza E Belluno, Cera Spaarbank, The Chugoku Bank, Limited, Copenhagen Handelsbank A.S., Credit Foncier de France, Credit Naval, Credito Romagnola, L'Europeenne de Banque, Fico France, Robert Fleming & Co. Limited, Fokus Bank A.S., Gulf Riyad Bank E.C., The Gunma Bank, Ltd., International Bankers Incorporated S.A., The Iyo Bank, Ltd., The Joyo Bank, Limited, Sal Oppenheim Jr et Cie, Postipankki, The 77 Bank Limited, The Shizuoka Bank, Ltd., Societe de Banque Occidentale, The Sumitomo Trust and Banking Co. Ltd., Trinkaus & Burkhardt (International) S.A., UBAE Arab German Bank SA, Deutsche Verkehrs-Kredit-Bank AG, The Bahraini

Kuwaiti Investment Group, The Bank of Kuwait and the Middle East K.S.C., Bankhaus Hermann Lampe Kommanditgesellschaft, Credit des Bergues, Banque Belgo-Zairoise SA-Belgolaise, Banque Intercontinentale Arabe, Banque Nordeurope S.A., Bank UCL S.A., FennoScandia Ltd., Forsta Sparbanken, OKOBANK, Saudi European Bank S.A., SKOPBANK, Volksdepositokas N.V. Savings-Bank

AGENT BANKS

National Westminster Bank PLC, Credit Lyonnais, Banque Nationale de Paris, Midland Bank plc

PAYING BANKS

International Westminster Bank PLC, Credit Lyonnais

CO-FINANCING AGREEMENT

European Investment Bank will finance up to the equivalent of £1,000 million as part of the total project facilities

4th November 1987

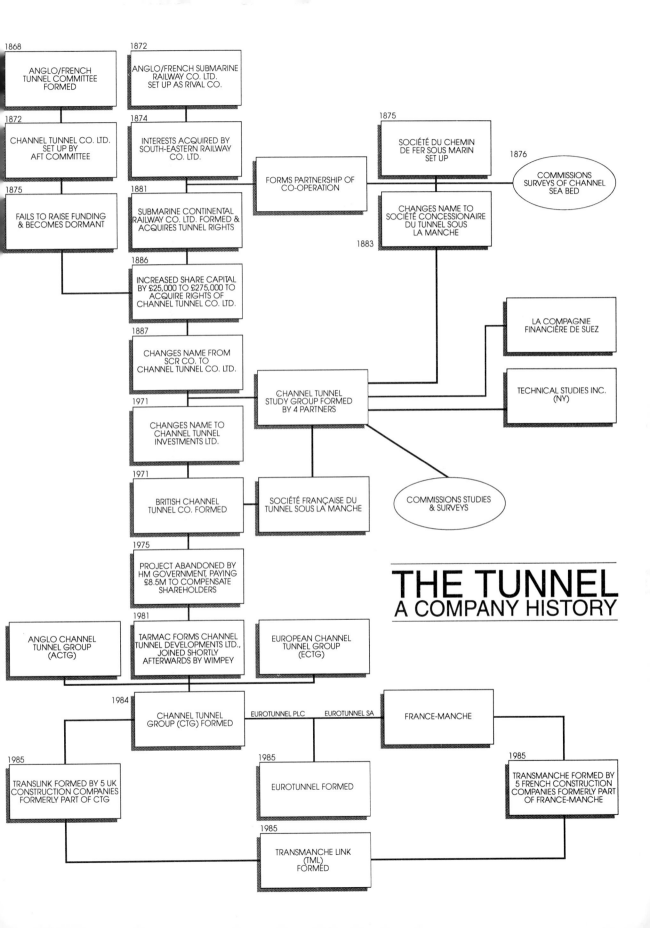

1868 — ANGLO/FRENCH TUNNEL COMMITTEE FORMED

1872 — ANGLO/FRENCH SUBMARINE RAILWAY CO. LTD. SET UP AS RIVAL CO.

1872 — CHANNEL TUNNEL CO. LTD. SET UP BY AFT COMMITTEE

1874 — INTERESTS ACQUIRED BY SOUTH-EASTERN RAILWAY CO. LTD.

1875 — SOCIÉTÉ DU CHEMIN DE FER SOUS MARIN SET UP

FORMS PARTNERSHIP OF CO-OPERATION

1876 — COMMISSIONS SURVEYS OF CHANNEL SEA BED

1875 — FAILS TO RAISE FUNDING & BECOMES DORMANT

1881 — SUBMARINE CONTINENTAL RAILWAY CO. LTD. FORMED & ACQUIRES TUNNEL RIGHTS

CHANGES NAME TO SOCIÉTÉ CONCESSIONAIRE DU TUNNEL SOUS LA MANCHE

1883

1886 — INCREASED SHARE CAPITAL BY £25,000 TO £275,000 TO ACQUIRE RIGHTS OF CHANNEL TUNNEL CO. LTD.

1887 — CHANGES NAME FROM SCR CO. TO CHANNEL TUNNEL CO. LTD.

CHANNEL TUNNEL STUDY GROUP FORMED BY 4 PARTNERS

LA COMPAGNIE FINANCIÈRE DE SUEZ

TECHNICAL STUDIES INC. (NY)

1971 — CHANGES NAME TO CHANNEL TUNNEL INVESTMENTS LTD.

1971 — BRITISH CHANNEL TUNNEL CO. FORMED

SOCIÉTÉ FRANÇAISE DU TUNNEL SOUS LA MANCHE

COMMISSIONS STUDIES & SURVEYS

1975 — PROJECT ABANDONED BY HM GOVERNMENT, PAYING £8.5M TO COMPENSATE SHAREHOLDERS

THE TUNNEL
A COMPANY HISTORY

ANGLO CHANNEL TUNNEL GROUP (ACTG)

1981 — TARMAC FORMS CHANNEL TUNNEL DEVELOPMENTS LTD., JOINED SHORTLY AFTERWARDS BY WIMPEY

EUROPEAN CHANNEL TUNNEL GROUP (ECTG)

1984 — CHANNEL TUNNEL GROUP (CTG) FORMED

EUROTUNNEL PLC — EUROTUNNEL SA — FRANCE-MANCHE

1985 — TRANSLINK FORMED BY 5 UK CONSTRUCTION COMPANIES FORMERLY PART OF CTG

1985 — EUROTUNNEL FORMED

1985 — TRANSMANCHE FORMED BY 5 FRENCH CONSTRUCTION COMPANIES FORMERLY PART OF FRANCE-MANCHE

1985 — TRANSMANCHE LINK (TML) FORMED

A Chronology
of Main Events

1707 The 'modern age' of tunnelling begins with the construction of the Urner Loch, on the St. Gotthard Road, Switzerland.

1785 The first crossing of the Channel by air in a balloon from Dover Castle to France.

1802 The Peace of Amiens marks the end of the Napoleonic Wars. Frenchman Albert Mathieu submits plans for a road tunnel between Britain and France to Napoleon Bonaparte.

1830 George Stephenson's steam-powered Rocket heralds the advent of railway travel.

1833 Frenchman Thomé de Gamond takes up Mathieu's plans and begins an investigation into the geology of the Channel, as well as into various schemes for a fixed cross-Channel link.

1856 De Gamond submits his research and proposals to Napoleon III. These are published the following year as *Étude pour l'avant-project d'un Tunnel sous-marin entre l'Angleterre et la France* (A study for an advance project for an undersea tunnel between England and France).

1860s Co-operation between the engineers William Low and John Hawkshaw expands into a full scale Anglo/French venture, subsequently to become, in 1868, the Anglo/French Channel Tunnel Committee.

1872 The Channel Tunnel Company Limited is incorporated and registered with the purpose of constructing 'an underground tunnel beneath the Straits of Dover, between England and France'. Rivalry develops between Low and Hawkshaw, and Low sets up his own Anglo/French Submarine Tunnel Company (later to become the South-Eastern Railway Company) in opposition.

1875 The *Société du Chemin de Fer Sous-marin entre la France et l'Angleterre* (the French Tunnel Company) is formed. The same year Colonel Edward Beaumont patents the first rotary boring machine.

1875 (August) The Channel Tunnel Company Limited receives Royal Assent to its Bill and a 99-year concession is granted.

1875 Preliminary works begin on the Channel Tunnel at St. Margaret's Bay, Dover. The site is later abandoned due to its physical unsuitability.

1880 The first shaft is sunk at Abbots Cliff, between Dover and Folkestone. Captain English's improved rotary boring machine, invented the same year, is used for the boring of the Tunnel.

1881 The second shaft is sunk at Shakespeare Cliff near Dover, and the French Tunnel Company sinks its first shaft at Sangatte near Calais.

1881 (August) The Board of Trade requests that the War Office set up a Commission to investigate the military implications of the Tunnel project.

1881 (December) Sir Edward Watkin forms a new limited company – the Submarine Continental Railway Company.

1882   (February) Political uncertainty in France and deteriorating Anglo/French relations give rise to fears of a French invasion. The Board of Trade intervenes on a technicality, forbidding construction work to continue.

1882   (July) British construction work stops.

1883   (March) The French abandon hope of Britain continuing with the project and orders its own construction-force to stop work.

1883   (April) A Joint Select Committee is appointed to investigate whether it is 'expedient for Parliament to sanction a submarine communication between England and France'. Three months later the Committee concludes that it is not expedient.

1886   (July) Watkin's Submarine Continental Railway Company buys the Channel Tunnel Company, and adopts its name.

1900   From the turn of the century until after the Second World War many schemes are proposed. Albert-Henri Sartiaux is the most prominent advocate.

1903   Wilbur and Orville Wright achieve the first motorised aeroplane flight.

1909   Louis Blériot becomes the first man to cross the Channel by aeroplane.

1931   A regular car-ferry service opens between Dover and Calais.

1955   The Rt. Hon Harold Macmillan, Minister of Defence, dispels British fears of invasion via a Tunnel, and thus opens the way for progress.

1957   The Channel Tunnel Study Group is commissioned to begin a geological and geophysical site investigation, as well as a full-scale investigation into various forms of fixed link.

1959   The hovercraft makes its first Channel crossing.

1961   The British Government announces its decision to apply for full membership of the European Common market. Shortly afterwards British and French Governments set up a joint working group to examine the proposal.

1966   (July) The British and French Prime Ministers make a joint announcement approving the construction of a Channel Tunnel.

1968   Civil unrest in France brings about a delay in proceedings.

1970   A new international consortium of banks, prepared to finance the Tunnel privately, is formed, and joins the British and French members of the Channel Tunnel Study Group.

1971   (July) Random voices of opposition in Kent become united to form the Channel Tunnel Opposition Association.

1971   (October) Parliament votes in favour of joining the EEC: Britain to become a fully-fledged member of the Community on 1st January, 1973.

1972   (October) British and French Governments and Tunnel companies sign an agreement covering the planned schedule of economic and technical studies and the construction programme.

1973   (September) British and French Governments announce that it is in their joint national interest to build the Channel Tunnel with a new high-speed rail link from London to Folkestone.

1973   (October) War erupts between Israel, Egypt and Syria.

1973   (November) The Channel Tunnel (Initial Finance) Bill is passed and receives Royal Assent. The Anglo/French Treaty is signed by British and French Foreign Secretaries to be ratified by 1st January, 1975. Preparatory works begin at Shakespeare Cliff and Sangatte.

1974   (February) War in the Middle East escalates into a major oil crisis. Prime Minister Edward Heath imposes a State of Emergency and, to conserve power, declares a three-day working week.

1975   (January) The British Government unilaterally abandons the Channel Tunnel on the grounds of economic uncertainty.

1979   (May) The Conservatives return to power under the premiership of Margaret Thatcher.

1981   (May) François Mitterrand is elected President of France.

1981   (July) Tarmac forms Channel Tunnel Developments (1981) Ltd. (CTD 81). Two months later CTD 81 is joined by George Wimpey in equal partnership.

1981   (September) François Mitterrand and Margaret Thatcher agree that a fixed cross-Channel Link demonstrates promise of considerable benefit to their respective national economies. Further studies are commissioned.

1982   (August) A New International Group is formed, comprising National Westminster and Midland Banks, Banque Nationale de Paris, Crédit Lyonnais and Banque Indo-Suez.

1984   (February) CTD 81 joins forces with the European Channel Tunnel Group (ECTG) and the Anglo Channel Tunnel Group (ACTG). The new consortium is to be known as The Channel Tunnel Group (CTG) and comprises five leading British Construction Companies.

1984   (November) President Mitterrand and Prime Minister Margaret Thatcher issue a joint statement: 'A fixed cross-Channel Link would be in the mutual interests of both countries.' An Anglo/French Working Party of Government Officials is appointed.

1985   (April) Flexilink is formed for the purpose of promoting and defending the role of the ferries on the short cross-Channel routes.

1985   (May) CTG enters discussions with the newly-formed French Group, France-Manche, comprising five major construction companies and banking associates.

1985   (July) France-Manche and The Channel Tunnel Group sign a Co-operation Agreement. This is to become a full merger and to form a new Anglo/French company: Transmanche-Link (TML).

1986   (February) President Mitterrand and Prime Minister Margaret Thatcher sign the Anglo/French Treaty, and a 55-year Concession is awarded. The Treaty is to be ratified the following year.

1986   (March) CTG opens an Information Centre at Tontine House, Folkestone.

1986   (June) France-Manche opens an Information Centre in the rue Mollien, Calais.

1986   (August) Eurotunnel and TML sign the Construction Contract – it is to be the beginning of a complex and turbulent financial relationship.

1986   (September) Equity 1 is launched, the first of three tranches to raise £1 billion of public equity, and is followed one month later by Equity 2.

1986   (October) Preliminary works begin at Sangatte

1987   (July) The Channel Tunnel Hybrid Bill reaches its Third Reading and passes into law as The Channel Tunnel Act. British and French Railways and Eurotunnel conclude The Channel Tunnel Usage Agreement.

1987   (July) Prime Minister Margaret Thatcher and President Mitterrand ratify the Franco/British Channel Fixed Link Treaty.

1987   (August) TML begins preparatory workings at Shakespeare Cliff.

1987   (November) Eurotunnel launches Equity 3. At a cost of £68 million, it is arguably the most expensive stock market flotation staged to date.

1987   (December) TML sinks the first shaft at Shakespeare Cliff.

1988   Militant opposition emerges in France over the routing of the new *TGV-Nord* line from Paris, via Lille, to the French Terminal.

1989   (March) BR secures an agreement with Customs and Excise in respect of 'on board' passport and customs' checks.

1989   (November) BR announces its choice of a private sector partner for the high-speed link: the Balfour Beatty/Trafalgar House consortium, later to become known as Eurorail.

1989   (December) The Intergovernmental Commission and the autonomous bi-national Safety Committee accept the principle of car drivers and passengers and coach passengers remaining in their vehicles during the shuttle train journey through the Tunnel.

1990   (January) TML completes one third of the total to be tunnelled for the entire project.

1990   (April) TML completes one half of the Channel Tunnel.

1990   (October) British and French tunnels meet. The official breakthrough ceremony is held one month later.

1992   (January) Britain becomes part of the European Economic Monetary Union.

1993   (January) The European Single Market Agreement comes into force.

1993   (July) Eurotunnel and TML sign an Agreement which ends their long-running financial dispute.

1993   (December) TML hands the project over to Eurotunnel.

1994   The Channel Tunnel opens.

Sources

I have cited many of my sources at appropriate points in the text. These include books, newspapers, journals and reports of various kinds. However, I give below a list of sources which I believe to be of particular importance in the Channel Tunnel story.

1. December 1968, Channel Tunnel requirements and possible location of Terminal facilities, Kent County Council.

2. 1973, *The Channel Tunnel Project*, Command 5256, HMSO.

3. May 1973, *The Channel Tunnel – Its Economic and Social impact on Kent*, DoE.

4. June 1973, *Channel Tunnel Economic and Financial Studies.*

5. May 1973, *A United Kingdom Transport and Cost Benefit Study*, HMSO.

6. April 1973, *The Channel Tunnel Project*, European Ferries Ltd.

7. June 1973, *The Channel Tunnel Economic and Financial Studies*, Coopers and Lybrand, *SETEC-Économie* Economist, The British Channel Tunnel Company and *Société Français du Tunnel Sous la Manche.*

8. July 1973, *The Channel Tunnel Project*, the Channel Tunnel Opposition Association.

9. July 1973, *Express Link with Europe – British Rail and the Channel Tunnel*, British Railways Board.

10. September 1973, *The Channel Tunnel*, HMSO.

11. January 1974, *Channel Tunnel, London – Tunnel New Rail Link*, British Railways Board.

12. February 1974, *Placing Memorandum*, British Channel Tunnel Company.

13. June 1974, *Channel Tunnel Rail Link*, a reappraisal of British Railways Board proposals, the Channel Tunnel Study Group.

14. September 1974, *Channel Tunnel Rail Link and alternative proposals*, Kent County Council.

15. October 1975, Submission to British and French Governments, Channel Tunnel Group.

16. 1980, *Cross Channel Rail Link*, British Railways Board.

17. June 1982, *UK/French Study Group Report*, Command 8561, HMSO.

18. May 1984, *Franco/British Channel Links Financing Group.*

19. April 1985, *Invitation to Promoters*, Department of Transport.

20. January 1986, *The Channel Tunnel Link*, Command 9735, HMSO.

21. 1986, Channel Tunnel Concession, British and French Governments.

22. 1986, Channel Tunnel Treaty; British and French Governments.

23. 1987, *The Channel Tunnel Story*, Michael Bonavia.

Index